CULTURE, PLACE, AND NATURE
Studies in Anthropology and Environment
K. Sivaramakrishnan, Series Editor

Centered in anthropology, the Culture, Place, and Nature series encompasses new interdisciplinary social science research on environmental issues, focusing on the intersection of culture, ecology, and politics in global, national, and local contexts. Contributors to the series view environmental knowledge and issues from the multiple and often conflicting perspectives of various cultural systems.

Caring for Glaciers

LAND, ANIMALS, AND HUMANITY
IN THE HIMALAYAS

Karine Gagné

UNIVERSITY OF WASHINGTON PRESS

Seattle

Caring for Glaciers was published with the assistance of a grant from the Naomi B. Pascal Editor's Endowment, supported through the generosity of Nancy Alvord, Dorothy and David Anthony, Janet and John Creighton, Patti Knowles, Katherine and Douglass Raff, Mary McLellan Williams, and other donors.

UNIVERSITY OF WASHINGTON PRESS
www.washington.edu/uwpress

Cover photograph of Himalayan village is by the author.
Interior photographs are by the author unless otherwise noted.
Maps are by Adam Bonnycastle.

LIBRARY OF CONGRESS CATALOGING-IN-PUBLICATION DATA
Names: Gagné, Karine, author.
Title: Caring for glaciers : land, animals, and humanity in the Himalayas / Karine Gagné.
Description: Seattle, Washington : University of Washington Press, [2018] | Series: Culture, place, and nature: studies in anthropology and environment | Includes bibliographical references and index. |
Identifiers: LCCN 2018011127 (print) | LCCN 2018031011 (ebook) | ISBN 9780295744025 (ebook) | ISBN 9780295744018 (hardcover : alk. paper) | ISBN 9780295744001 (pbk. : alk. paper)
Subjects: LCSH: Glaciers—Himalaya Mountains Region. | Glaciers—India—Ladakh. | Human ecology. | Environmentalism.
Classification: LCC GB2559.H56 (ebook) | LCC GB2559.H56 G34 2018 (print) | DDC 179/.1—dc23
LC record available at https://lccn.loc.gov/2018011127

To my daughter, Nilza, who came into this world on a winter day so cold it broke Canadian temperature records, and who has since brought warm sunshine into every single day of my life.

CONTENTS

FOREWORD

AS ENVIRONMENTAL ANTHROPOLOGY GRAPPLES WITH THE IDEA OF living in the Anthropocene, scholars have expectedly taken different paths to finding what constitutes a credible object of study and how it must be framed in philosophical terms. For many, the gravity of the realizations forced on those contemplating life in the Anthropocene has provoked a rethinking of the ontological status of humanity. This has given rise to anthropological approaches that variously identify as posthumanism, the anthropology of life, new materialism, or anthropological examination of the more-than-human world.

These discussions are not entirely new, and some will suggest they can be dated at least in recent times to the writings of Donna Haraway, Tim Ingold, and Bruno Latour.[1] Others will aver that these ideas also build on the persistent effort in branches of environmental anthropology to quiz the Cartesian nature-culture divide that many feel has contributed to the accelerated processes of environmental destruction that have in turn created the Anthropocene.[2] Climate change in this view is but the most dramatic and widely felt consequence of human disconnection from nature and the increasing inability to see human and other life, as well as that which may be lifeless, as inextricably woven together on this planet. In this view, all things share a common fate—they will be destroyed, largely by the egregious misconduct and greed of the human race.

Apart from a debate on the best theoretical or philosophical orientations for studying these world-changing processes, especially in the humanities and interpretive social sciences, there is also increasing disquiet in some quarters with the fact that while only some humans (very few as a proportion of the world population) created the problems, the ones least responsible face the worst effects. These concerns have given rise to a subfocus within environmental justice scholarship on climate justice. So, if humans are seen as one race, or species, brutalizing the earth, such a view obscures the reality of huge discrepancies that have emerged among humans in power and wealth over modern historical time.[3] We might well ask if the concept of the Anthropocene masks the magnification of environmental injustice, not just toward the nonhuman world, but also toward those who are paying a heavy price for the excesses and hubris of a small minority who have lived far from (or can escape from) where the vast majority are now struggling with their fate.

These are the issues raised evocatively in this vivid examination of Ladakhi life in the Indian Himalaya. Drawing on knowledge gained during arduous and sustained fieldwork in the high mountain villages of Ladakh, living through the biting winter with her interlocutors in their pastoral communities, Karine Gagné attends to regional history, senses of place, and social practices of ethical conduct. She argues that how Ladakhis think about their land, the climate, the retreating glacier, and their profound losses is informed by both their Buddhist heritage and a practical ethics that stems from a deep attachment to, and respect for, their land and its gifts—a land they feel they have betrayed by straying from the moral compact that shaped their own lives and the condition of pasture, glacier, and farm, as well as animals and air, and the meaning of time, seasons, and fluctuating temperatures.

Gagné grounds this anthropological study of living in the Anthropocene in the midst of climate change as it unfolds across a rural landscape transformed by economic development and borderland military activity. She shows how moral and ethical concerns animate the natural world. Equally she shows that the biggest environmental crises of the day remain locally experienced as both physical and cultural phenomena, shaping conduct and values and changing livelihoods and sentiments. Much of Gagné's work is done with elders who still live in villages, since most of the younger generation are urbanized or have become migrants in search of education and jobs outside Ladakhi villages or even Ladakh itself. Partition and mid-twentieth-century wars, which generate an affect-laden relationship with the Indian nation mediated through its military, come to vitiate an ethical relationship with land, animals, and glaciers.

War and militarism generate both moral ambiguity and economic opportunity, resulting in transformation of Ladakh that is cultural, material, ethical, and practical. These changes in turn affect the moral compact between Ladakhis and their environment in relation to farming, pastoralism, and the retreating glacier that is the greatest cause for concern for Ladakhis and climate scientists alike. This strikingly original work of environmental anthropology on climate change and the Anthropocene remains grounded in local history and values and the ways in which the local and the global destabilize each other though critical events and subtle shifts in affective ties and the moral foundations of interspecies relations.

K. SIVARAMAKRISHNAN
YALE UNIVERSITY

PREFACE

THE RESEARCH AT THE HEART OF THIS BOOK BEGAN BEFORE I FIRST set foot in Ladakh. My initial contact with Tibetan Buddhist culture and societies took place in 2004 when I traveled for the first time to Dharamsala, a sprawling town in Himachal Pradesh state in North India where I eventually spent sixteen months working for human rights NGOs. As the seat of the Tibetan government in exile and the residence of the Dalai Lama, the town has a large population of Tibetan refugees. The time I spent living and working among Tibetan friends and colleagues in India and elsewhere has taught me much about Tibetan Buddhist culture and philosophy, and has formed me both as a researcher and, I believe, as an individual. One of the most significant aspects of my friendship with Tibetan refugees while in Dharamsala were the endless nights spent listening to them recount life stories and fables from Tibet. They would inevitably break into song, taking turns chanting a cappella the lines they had once sung while tending to their grazing yaks in Tibet. And I was always left mesmerized: I seemed to have fallen into a group of friends collectively gifted with beautiful singing voices. Their repertoire of folk stories and songs was deeply infused with notions of morality and abounded in environmental metaphors and references, which required much effort on my part as, still a student of the Tibetan language, I tried to grasp the implicit nuances. In this way, I came to learn that figures of speech, proverbs, and analogies that invoke environmental imagery are idiomatic for Tibetans.

What I gleaned from this experience is that morality does not pertain to doctrine only. While Buddhist philosophy may at some point have informed or inspired these narratives and songs, they are also the product of their context. For instance, the environmental referents of these songs and the fact that they emphasize love for the grassy hillocks and the animals reflect an attachment that herders have nurtured through their long experience of life in the mountains. As one former herder told me when I asked him why their songs and moral stories so often refer to elements of the natural world, "This is what we are surrounded by." His answer reveals the way in which the crafting of people's affective and moral universe is based on life experience. The fact that these songs may be sung out of nostalgia for a landscape that the refugees would perhaps never see again also illustrates, as anthropologists have long demonstrated, that folk repertoire and its meaning should be considered in light of specific social and historical contexts. For instance, my friends would frequently sing a song about a herder who misses his mother while in the mountains and who feels guilty for not being able to take care of her while he's away. But as sung today, it is clear that the mother in the song stands for the land of Tibet itself and for those who remain behind.

This window into Tibetan Buddhist morality departs from the common means of understanding Tibetan Buddhist culture and community life, which is through a focus on religious doctrine. For many Tibetan studies scholars, an appraisal of the esoteric teachings and contemplative practices of Buddhism is key to understanding how Tibetan Buddhists think and act (Childs 2004:1–2; Kapstein 2013:3). Although I concur with their perspective to a degree, I also believe that emphasizing the doctrinal dimensions of religion alone produces an oversimplified perspective that "cast[s] Tibetans as a people incarcerated in an eternal past" (Childs 2004:1–2). If Buddhist philosophy exerts a deep influence on lay Ladakhis, their actions and reflections are also the product of secular cultural values. As anthropologist Charles Ramble, arguing against a tendency to adopt "bookish standards," explains: "Familiarity with Buddhist literature equips the researcher to understand Buddhism; by itself it cannot enable him to know how the religion is incorporated into the tradition of a particular community" (1990:185–86, cited in Childs 2004:1).

To think of Ladakhis' sense of morality as being the product only of doctrine and religious teaching is limiting. Not only would it be presumptuous to assume that all Ladakhis are well versed in doctrinal religious matters, but it would also ignore other means through which notions of morality and ethical selfhood are cultivated. Are folk stories and songs less valuable than doctrine in trying to understand the worldview of some Tibetan Buddhists and their conceptions of morality? Is the moralization of the natural environment by Ladakhis less valuable as a window into their culture because it belongs to popular forms of religion or because it is cultivated outside a religious context? To assume that the way people craft their moral universe needs to be validated against doctrinal religion for it to be valuable would be unfortunate. Furthermore, to think of religion as the only moderator of people's life would be to consider notions of ethics and morality in isolation of social change. As is true of any religion, adherents of Buddhism experience the universal tension between individual aspirations and cultural ideals (Childs 2004:2–3), a tension that is particularly salient in Ladakhi society today. Indeed, a sense of morality and notions of ethical selfhood may take on particularly strong meaning when a society experiences profound transformations.

My aim is not to produce a work on Buddhist religious philosophy, nor to study morality as embodied in prayers, religious discourses, or precepts. Scholarship on the Himalayas and elsewhere has already demonstrated the role these practices play in efforts to lead an ethical life. In this book, rather, I attempt to show how, for many Ladakhis, mundane activities such as farming and herding are equally important ethical practices in forming the basis for living a morally correct life. I also hope to demonstrate that Ladakhis draw their moral bearing from a variety of sources, whether from folk rituals and beliefs or practical engagement with the environment. For instance, their love of animals and their desire to care for them in a world where livestock herds are being increasingly reduced in size is the result not just of a religious creed, but of attachment developed through herding in the mountains. This book primarily takes the perspective of the life of ordinary, elderly farmers and herders. Although the voice reflected here is not that of religious experts, these elders nonetheless see the environment as suffused with moral meanings. Thus, although I do at times focus on rituals and discuss elements of Buddhism, I do so with the intention of understanding what this says about people's relationship with the environment and how this relationship is changing today.

Let me first start with a note on Ladakhis' pragmatic approach to religion. For many, performing prayers and rituals is a way to achieve prosperity, to avoid misfortune, and to respond to the contingencies of everyday life. Scholar of Buddhism Matthew Kapstein (2013:3) observes that the concerns that drive daily religious life in Tibetan Buddhist communities are those inherent to preserving, rather than transcending, the order of the world: ensuring that the harvest is abundant, that the livestock remain healthy, that diseases are kept at bay. Kapstein notes that such considerations pertain to the "facts on the ground" of Tibetan Buddhist religious life. This constitutes a very accurate portrait of Ladakhis' relationship with religion. In fact, when inquiring with religious experts (monks, *geshes*, high lamas) about specific prayers recited during rituals, I was repeatedly told that lay people "have no idea about the meaning of the text." Although one should be careful not to generalize this observation to all lay Ladakhis, what such a statement means is that even though Ladakhis may contribute money toward rituals or attend community prayers, what interests them is the end result rather than being well versed in a spiritual discipline.

My first encounter with this reality occurred the very first day I arrived in Ladakh during my preliminary research stay, in 2011. For several days, three monks read prayers all day long in the kitchen of my host family, while the various family members carried on with their daily activities. What intrigued me was the fact that none of them participated directly in the prayers, and this did not seem problematic for the ritual's intended outcome. But this is also part of the pragmatic orientation of Tibetan Buddhism: religious experts are the mediators between the laity and deities, and one can call on their services to obtain favors without having any knowledge of the scriptures.

KNOWLEDGE: COPRODUCTION AND SITUATEDNESS

The research in this book draws from a two-month preliminary research stay in Ladakh in 2011, a twelve-month continuous stay between 2012 and 2013, and, to some extent, a six-month stay in 2016. The bulk of the fieldwork on which this book is based would not have been possible without the help of my diligent research assistant, Namgyal, with whom I worked throughout the my 2012–13 stay. Namgyal accompanied me to every village I visited to conduct interviews and gather data. Crucially, it was he, as well, who introduced me to Ladakhis of a younger generation than those I met in villages. As I explain in the introduction, this book takes the perspective of my main interlocutors in the field, namely aging farmers and herders. Although the voice of young adults is overshadowed by the voice of elders in this study, my informal discussions with younger Ladakhis about their lives and aspirations provided me with a key vantage point from which to examine the generation gap that is so present in today's Ladakh and that occupy significant portions of my inquiry. Namgyal is a recurring figure in this book, and the concerns and challenges he is experiencing are those faced by many Ladakhis of his generation. These challenges recur throughout the book: in chapter 2, the fragility of his individual identity as a result of years of schooling outside Ladakh is exposed during an event in which he comes across as an outsider of his community; in chapter 3, Namgyal is troubled by his moral duty to provide farming help to his father, who is away for military work, despite the fact that he has felt neglected by his father for most of his life; and chapters

4 through 6 demonstrate Namgyal's lack of familiarity with the habitus of the agrarian and pastoralist worlds.

Questioning the long-held convention in academia to leave the contribution of research assistants unstated, recent views argue that we recognize the social nature of the "field" and the role played by research assistants in its emergence through ethnographic practices (Middleton and Cons 2014). For the field is never a given space that simply lies there waiting for the anthropologist to come scrutinize it, but rather is "constituted by the network of connections and linkages forged in doing fieldwork" (Gupta 2014:399). Research assistants often play a key role in the creation of these networks, and therefore in how "the field" comes into being. Witty, passionate, and curious, Namgyal took to fieldwork with gusto and was unmistakably enthusiastic about contributing to my research project. If he remained throughout my fieldwork an invaluable resource, helping me to develop the connections that led to this book, the most salient aspect of his contribution was undeniably his capacity to forge an emotional connection with our elderly interlocutors.

At the outset, I had not anticipated that conducting research on perceptions of environmental change would draw so poignantly on such sensitive issues as the loneliness of elderly people in the villages and their desperation at having to sell their precious livestock because there is no one left to maintain the herds. Elders feel they are increasingly left alone in villages, but as we shall see, the causes of their feelings of abandonment are varied and complex. New education and work opportunities have resulted in an intermittent, rather than permanent, presence of the younger generations in villages. Although no villages are inhabited exclusively by elders and younger people are still present, their engagement with community life has changed. The fabric of communal life, as elders once knew it, is increasingly fragmented. As a result, many rituals, gatherings, festivals, and other community practices are waning or have been completely abandoned. When speaking of the increasing degree of anonymity in village life, elders told me on numerous occasions that "today, we don't even know when a child is born," alluding to elaborate ceremonies that used to mark the birth of a child but that today are increasingly truncated or neglected altogether.

One particular event that occurred during fieldwork made me even more conscious of the stigma associated with the situation of elders. At one point, I admitted to a Ladakhi friend my sadness over elders' confessions of loneliness. To my surprise, he reprimanded me harshly for "judging Ladakhi society by Western standards." Elders in Ladakh, he explained, unlike those in the West who are so often placed in old-age homes and left in the care of others, are well taken care of *when they need it,* whether it is for health or other reasons. Yet it is precisely the conditional character of this attention that is so difficult for Ladakhi elders. They miss the everyday presence of younger family members and neighbors and the lively village life they once knew. Moreover, although deference to and respect for seniors continue to be valued in theory, in practice the clash between social changes and traditional life has contributed to the destabilization of certain relationship ideals between generations. My friend was angered by my suggestion that his fellow Ladakhis were neglecting the moral duty of providing for elders. But what elders in the villages were confiding, sometimes directly, sometimes indirectly through the use of rhetorical processes, suggested exactly that. Elders often expressed this as the feeling of having become useless in society or even of having become a burden—and in its most extreme form, as a desire for death.[1]

I quickly realized that Namgyal's presence played a significant role in elders' openness toward us on these issues. I believe that this had much to do with the rapid social changes taking place in Ladakh in recent decades. Many elders are worried about what Ladakhi society will become. Younger generations, in their view, are more interested in cosmopolitan life than in village life. As a result, my interlocutors were enthusiastic about seeing a young Ladakhi interested in their perspective on local history. In addition, being in the latter stages of their lives, many elders felt that some parts of the local history needed to be exposed. For many elders whose voices have lost a measure of influence in the wake of recent social changes, the interviews had a cathartic effect. When answering a simple question, their words often came pouring out, as if they had waited years to share their views on issues affecting Ladakh.

But more than anything, the success of our interview sessions had much to do with Namgyal's life trajectory. From the start, his life had not been easy. Due to tragic life circumstances, at the young age of three he was sent to a monastic school established by a Ladakhi monk in South India. This was as far away from his native mountains as could be, especially since road conditions at that time made travel all but impossible. After a few years living in the monastery, where hunger prevailed, Namgyal and a few other Ladakhi children were adopted by an Indian couple and educated in a school system founded on the nontraditional views of a world-renowned Indian philosopher. In the following years, he frequently moved around the country, constantly experiencing rounds of bonding and separation. Although he made a few sporadic trips to Ladakh during this period, it was only around the age of twenty, about a decade before our work together, that Namgyal returned to live in Leh. Yet despite his having spent many years in what to a native of the mountains was a foreign land, his sense of belonging to Ladakh remained intact.

Still, the return to his homeland was not easy. For one thing, although a brilliant student, as his friends insisted, Namgyal was not assiduous at studying or showing up for class. He ended up failing his board exams at the end of his secondary education, which prevents him from securing employment in the civil service. As a result, he had had different jobs in the tourist industry, an occupation he had recently started to limit to monastery tours, given his increasing unwillingness to go trekking in the mountains due to altitude sickness. This malaise was perhaps created by the toll on his body of heavy cigarette smoking, his medicine to deal with the stress of his homecoming.

Namgyal lived with his mother, though for years he felt like a stranger in her house. A stressful aspect of his life had been the court case he had initiated for his mother, who had lost, to her elder sister, her share of land inheritance when their father passed, on account of primogeniture. Having been educated in a progressive schooling system, Namgyal was well aware of the law in matters related to land inheritance. In a place like Ladakh, where gossip acts as a powerful form of social control, the situation also had its toll on the family. Although the case was resolved in Namgyal's mother's favor relatively quickly—especially considering the snail's pace of court procedures in India—it nonetheless consumed Namgyal's time and energy for several years. He was in fact still buried in related administrative papers when I started to work with him.

Yet despite its trials, Namgyal's distinctive life trajectory has endowed him with a profound humanism that translates into an exceptional ability to connect with vulnerable people. Perhaps because he spent so many years outside the region, Namgyal also developed a perceptive curiosity for the details of local history, coupled with an objectivity toward the many dimensions of Ladakhi life. I believe that Namgyal, not only because he was a local figure but also because he was an attentive and passionate listener, was instrumental in making elders feel comfortable—indeed enthusiastic—about discussing sometimes highly emotional and intimate subjects. For these reasons, I see the fieldwork component of my research as the result of teamwork, rather than as the product of one anthropologist's lonely craft.

Yet if fieldwork is indisputably about teamwork, it is also a highly personal experience, as the methodological tools used by the anthropologist are obviously entangled with the anthropologist's self. The researcher who strives toward objectivity must nevertheless acknowledge her involvement in the context under study. Perhaps out of a desire for transparency, it is only instinctive for me not to mute my voice, and it will soon become clear to the reader that my voice, not only as a researcher, but also as an individual experiencing Ladakh, is quite present in the pages that follow. In many ways, my subjective experience of the field enabled me to tackle a range of issues. As an outsider not fully adapted to Ladakh's environment, I was also at times vulnerable in the field, both physically and psychologically. My experience of the challenges of life in this mountain environment was crucial to shaping my understanding of the difficulties that Ladakhi elders themselves face, and made me especially sensitive to their perspective, perhaps to the neglect of other facets of Ladakhi life that a different researcher might well focus on. In the end, I hope that the account thus produced may serve as a reminder that knowledge is never just information or data; rather, knowledge is always situated, often coproduced, and deeply relational, and its presentation, ultimately, is inescapably human.

ACKNOWLEDGMENTS

THE DEBT ACCRUED ALONG THE PATH THAT LED TO THE COMPLE-tion of this book is large and I will probably not be able to repay it in this life-time. First, I wish to express my unbounded gratitude to the inhabitants of Sham for hosting me and for generously and kindly taking the time, through words and guidance, to allow me the opportunity to become immersed in the everyday life, history, and memories of this part of Ladakh. I extend as well my deepest thanks to my research assistant and collaborator, who per his request has been renamed Namgyal in the pages that follow—for trek-king kilometers and kilometers with me in the mountains in search of his-tories embedded in the landscape, for carrying my bags when I did not have the strength, for taking care of me when I was sick (which was often), for loyal help with everything during my fieldwork, for being an incomparable guide: in short, for being his unique and colorful self. And I sincerely thank Namgyal's family for the generous place they made for me in their life.

My work in Ladakh would not have been the same without the friendship and support of Amit, Anjeli, Dorjey, Gulzar, Gurmet, Odzer, Rigzin, Sonia, Stanzin, Thundup, and Tsewang. Many people in India, locals and others, from all walks of life, including scholars, artisans of Ladakhi society, and many others, offered invaluable help at various stages of my fieldwork. I am grateful to Richard Hendin, Blaise Humbert-Droz, Geshe Khampo Konchok Phandey, Geshe Konchok Namgyal, Morup Namgyal, Tashi Rabgias, Hank Tomas, and Rebbeca Norman and Sonam Wangchuk from the Students' Educational and Cultural Movement of Ladakh (SECMOL). I am also grate-ful to Sonam Wangchuk of the Himlayan Cultural Heritage Foundation for his generosity.

My accounting of the many debts accumulated on this journey would not be complete without acknowledging its very beginnings, in Dharamsala in 2004. *Thugs rje che* to Kalsang, Lobsang, Ludhup, Namlha, Norgey, Pema,

Raja, Samdup, Tinley, Tsultrum,and my dear friend Zorgy. You have all been such wonderful company through my stays in India. You sowed the seeds of this project. I am grateful that our paths crossed at one point in (this) life. Thank you for looking out for me through the years and for treating me like kin.

A number of scholars are owed thanks for taking the time to share their expertise on Ladakh in the course of my research, in particular Fernanda Pirie, John Bray, Katherine Hay, Martjin van Beek, Noe Dinnerstein, Ravina Aggarwal, and Seb Mankelow. I am also grateful to Harjit Singh for welcoming me as a research scholar at the Centre for the Study of Regional Development at Jawaharlal Nehru University. For their help with my numerous questions, my most sincere thanks go to Daniela Luschin-Wangail, Jigmet Lhundup, and Tashi Wangail Sothi.

Bernard Bernier, Christophe Jaffrelot, Galen Murton, Georges Dreyfus, Gilles Bibeau, Gordon Bromley, Isabelle Henrion-Dourcy, Jan Salick, John Leavitt, Katia Buffetrille, Kenneth Hewitt, Lara Braitstein, Marc Des Jardins, and Robert Moseley have generously helped me with various aspects of my research. Adam Bonnycastle has been supportive with the creation of the maps for this book.

Thanks also go to Ben Orlove, Guy Lanoue, Karine Bates, Kregg Hetherington, and Mattias Borg Rasmussen, who read this manuscript at various stages and provided insightful feedback. Maciek Janicki took on the task of meticulously revising my words at one phase of this project, a laborious endeavor that Karen Caruana undertook when it came the time to revise this manuscript. Both were amazing in improving the quality of this manuscript. At University of Washington Press, Lorri Hagman provided valuable editorial comments and both she and Julie Van Pelt were excellent guides.

Karine Bates has been a steadfast source of advice and provided wise guidance during the many years it took for the completion of this project. Always encouraging, Ben Orlove has been a great reader and source of advice as well, and I am grateful for his support over the years and for providing me with an amazing learning environment during my residence at Columbia University.

I was lucky enough to have a wonderful mentor and reader at a crucial moment in this project. K. Sivaramakrishnan ("Shivi"), always encouraging and generous with his time, provided me with the perfect environment at Yale University, where most of this book was written. There I had the chance to meet people who read and commented on parts of the text: Alder Keleman, Amy Johnson, Amy Zhang, Aniket Aga, Bhoomika Joshi, Daniel Tubb,

Deepti Chatti, Elliott Prasse-Freeman, Luisa Cortesi, Meredith Mclaughlin, Sahana Ghosh, and Souvanik Mullick.

Parts of chapters 2 and 3 were presented during a workshop on Territoriality in December 2014 at the University of Copenhagen. I am thankful to Christian Lund and Mattias Borg Rasmussen at the Department of Food and Resources Economics at the University of Copenhagen for inviting me to participate in this workshop. I am also thankful to Evan Berry and Robert Albro for inviting me to participate in the American University's Center for Latin American and Latino Studies workshop on Religion and Climate Change in Cross-Regional Perspective in Peru, where chapter 4 was presented. Both avenues were an opportunity to receive incisive comments on my work.

I am lucky to come from a country where access to education is by and large democratic. I am extremely thankful to Canadian taxpayers for their generous support that made possible two fellowships from the Social Sciences and Humanities Research Council (SSHRC). Both the SSHRC's Banting Fellowship and a Joseph-Armand Bombardier Scholarship were crucial to the completion of this project over the years. I am also grateful for the financial support I received from a variety of institutional sources, including the Canadian Federation of University Women; Les Offices jeunesse internationaux du Québec; la Faculté des Arts et des Sciences de l'Université de Montréal and the Faculté des Études Supérieures et postdoctorales de l'Université de Montréal; the Shastri Indo-Canadian Institute; the International Development Research Centre; and the Canadian Anthropology Society.

It has been a long journey, often amazing and at times challenging. Nilza, my daughter, came to this world when I was writing this book and she provided so much love and joy. This project would not have come to fruition without the love, sacrifices, and support, especially during my most trying moments, of my parents, Carmen Labelle Gagné and Richard Gagné. Thank you.

NOTE ON TRANSLITERATION

LADAKHI IS A TIBETAN LANGUAGE, AND LADAKHI AND TIBETAN have a common writing system. Yet individual words can differ markedly in pronunciation—not just between Ladakhi and Tibetan, but also across Ladakh itself, such that the way the people of Sham pronounce a word may differ from its pronunciation in Leh or in the Nubra Valley. In this book I render Ladakhi words as they are spoken in the Sham dialect, transcribed here as they would be pronounced by English speakers. At first mention of a word, I also provide, where known, the standard Wylie (1959) transliteration, in parentheses (unless the two pronunciations are identical). For example, the term for a local astrologer is rendered *onpo* (*dbon po*) at its first instance. Thereafter, I simply use *onpo*. To make the text less cumbersome, I do not provide the Wylie transliteration for longer expressions or for the names of buildings. The glossary includes the Ladakhi words that appear in this book, their Tibetan form, and an English translation. Words in Hindi and Urdu are noted as such.

When transliterating Ladakhi names, I use the most common transliteration: for example, Dorjey instead of *Rdo rje*. To maintain the privacy of my interlocutors, I use pseudonyms throughout the book. I have also changed the house names, a common form of identification in Ladakh. Ladakhis, like Tibetans, generally do not have surnames, but rather two given names chosen by a Rinpoche (incarnate lama) soon after their birth. When talking about someone, people may use the person's two names or the first name only, sometimes preceding it with a designation such as *abi* (grandmother) or *meme* (grandfather). The *le* added to a term of address—for example, *abi-le* or *meme-le*—is a polite particle indicating respect.

CARING FOR GLACIERS

Introduction

Morality and an Ethics of Care in the Himalayas

AFTER A LONG WINTER, SPRING HAS FINALLY COME TO LADAKH. Everywhere in Sham,[1] people are leaving the comfort of their kitchens and the warm fires of their wood stoves as the temperature outside becomes more and more agreeable. The blanket of snow covering the surrounding summits is slowly melting, the precious meltwater feeding the streams that irrigate terraced fields through ingeniously complex arrays of channels. Soon, the glacier meltwater will quicken the current of the streams on which the region's agrarian village economy depends.

We are in Ang, in the upper part of Tingmosgang village. *Abi (i pi)*[2] Lobsang is preparing *kahwah* tea for us, a traditional infusion from Kashmir Valley. In the pot, she brings to a boil water with tea leaves, cinnamon bark, cardamom pods, and sugar. Then she adds her special ingredient, a handful of apricot kernels, the precious delicacy of Sham. A pleasant fragrance fills the room, where walls are covered with soot left by smoke from the fireplace over the long winter months. *Abi* Lobsang hands me a cup. With its abundant quantity of sugar, typical of Indian tea, and the fragrant spices of the Kashmir Valley, my cup of tea stands as a microcosm of the diverse influences that have traveled across the high summits of the Himalayas to our Buddhist Ladakhi host's kitchen.

But all is not well with *abi* Lobsang. A few weeks earlier, a snow leopard broke into her animal pen, right beside her house. Nine sheep and goats fell easy prey to the big cat. Two ox calves survived the carnage, but one of them died a while later. The incident deeply upset *abi* Lobsang. Upon hearing the news, villagers had gathered at the elderly woman's house. In the meantime, she had managed to trap the snow leopard in a shed—an easy task, she says humbly, as the beast was "drunk on blood." Some of the village men wanted

3

I.1   Villagers looking at a snow leopard captured after an attack.
Photo by Jigmet Lhundup.

to kill the animal, but *abi* Lobsang stopped them. It happened to be an auspicious day in the Tibetan calendar and it would have been quite inappropriate
to kill a living being.

This was a few weeks ago, but *abi* Lobsang is still in shock. She feels sorry
for the animals she was rearing in very difficult conditions. At seventy-five
years of age, she has found it challenging to take care of the animals. But
despite the shortage of help in her household, she is still resisting selling her
livestock. She fears that letting them go to Muslim buyers, as many villagers
do in Sham today, will mean the slaughterhouse for her sheep and goats. Why
her, she wonders? Why such a retribution? She is living an honest life, making regular offerings at the monastery, fulfilling her community duties, and
participating in rituals and festivities. But her children are working outside
the village, as is true of many Ladakhi households now, and have not taken
up work on the farm. She wonders if they ever will. She is struggling with
the effects of old age and the work necessary to maintain the farm.

Soon after the loss of her animals, other villagers invited a local *rinpoche*,
an incarnate lama, to visit the village. The Rinpoche told them that the event
must have been a form of retribution against the community. His words
provided *abi* Lobsang with a measure of solace, for they meant she was not

the sole culprit behind this unfortunate event. But she still misses her animals dearly.

Misfortune seems to befall the community of Ang with particular insistence this spring. This is the sowing season and the whole hamlet is short on water to irrigate the fields. There is less water in spring because there is less snow these days, *abi* Lobsang explains. "Why is there less snow?" I ask. The old woman pauses for a moment, pensive. With her arthritic fingers, she rolls the beads of her rosary, searching for an answer. "It could be due to the warmer temperature," she suggests. "It is much warmer in winter these days." She then turns to the large turquoise tin thermos, ornamented with a painting of a cherry blossom tree, and refreshes my cup of tea. The warmth of the mug provides relief for my cold fingers. Here, at an altitude of 3,600 meters, spring still feels cold after a long winter.

My question having piqued her curiosity, *abi* Lobsang carries on with her observations. The water in the stream that passes through the village was in the past much higher in springtime, she says, a change she attributes to the recession of the local glacier and decreased snowfall. *Abi* Lobsang then recollects how, when she was young, villagers would bring charcoal into the mountains in order to grow Kangri Soma (Gangs Ri So Ma) the "new glacier," which stands next to the considerably larger "old glacier," Kangri Nyingpa (Rnying Pa). They would do this "when there was a fear that the glacier would go," she explains. As a young girl, she had found it a pleasant activity to go around with other villagers collecting charcoal from the fireplaces of every household. This was a community responsibility. Everyone had to provide charcoal for Kangri Soma. "This is what people would do," she says, reminiscing; "they would bring charcoal to the mountains and throw it on the glacier to make it grow."

On numerous occasions in Sham, I heard elders give such quasi-oneiric accounts of villagers trekking to glaciers with bags full of charcoal. All indications are that this practice now belongs to the past. Elders saw it performed in their childhood, but not since.[3] "Why aren't people doing this today to try to help with the water problems?" I ask. "I don't know," she says, amused. "Things were different before; now people have become really empty at heart." The implication of a changing attitude for the state of the environment is a leitmotif that suffuses the discourse of Ladakhi elders. If this reflects the entanglement that Ladakhis see between humans and glaciers, it also prompts the question: How is it that the prospect of losing glaciers compelled community mobilization in the past, whereas today it fails to elicit a similar reaction?[4] Whether or not such practices were effective in growing glaciers and maintaining water supplies is not what concerns us here. Importantly for this discussion, however, *abi* Lobsang's account reflects a change in people's relationship with the environment and a changing ethics of care for glaciers.

After the independence of India, successive wars with both China and Pakistan, and the decentering of the agro-pastoralist economy, Ladakh became a sensitive border area. The resulting military production of the Indian state brought the economic restructuring of the region. The ensuing work opportunities in military service and employment related to the military, along with the ever-increasing presence of state institutions in the region, all contribute to significant demographic changes, characterized especially by rural depopulation, and have precipitated major social changes. Village life is no longer the same, the emotional impacts of which a growing number of elders like *abi* Lobsang are experiencing through their isolation.

In parallel to changes brought by state formation, the landscape of Ladakh is being transformed by climate change, which is evident in glacier recession and environmental degradation. Elders see the root causes of this shifting landscape in a changing moral order that no longer values the reciprocity between human actions and the condition of the natural world. For many, this is apparent in the abandonment of village life, the end of community rituals and responsibilities, the rise of individualism, their own increasing isolation, the alienation of youth from rurality, and the widespread disinterest in agrarian and pastoralist work. As *abi* Lobsang implies when she explains that people no longer care for glaciers, this changing moral order is also evident in the way people relate with the environment.

AN ETHICS OF CARE

The ethics of care that has long existed for the nonhuman world in Ladakh—in particular glaciers, the land, and animals—has been transformed under a politics of nature deployed by the Indian state that aims to reconfigure Ladakh into a mountain fortress for India. This ethics of care is not the exclusive purview of religious doctrine, but exists in a variety of practices and narratives and is nurtured through engagement with the environment.[5]

This emphasis on ethics as practice builds on recent discussions in anthropology inspired by Michel Foucault's later writings.[6] Rather than a set of rules and principles that pertain solely to texts and discourses, ethics, as an element of everyday life anchored in specific practices, can be viewed as a "salient dimension of human activity and of human being in the world" (Lambek 2015: chapter 1). Moreover, by considering ethics and morality as objects of study in their own right, anthropologists are moving away from a long-prevailing approach that conflates morality and ethics with culture, discourse, and ideology, a perspective associated with the collective sociology of Durkheim (Laidlaw 2014). Yet this is not to say that ethics and morality can be isolated from broader structures, because an anthropological approach to ethics as an orientation to living "the best life possible" requires an appreciation of "the kinds of worlds in which people do live," worlds that are defined in part by religion, culture, and other dynamic structures and circumstances (Lambek 2015: chapter 1).

The ethics examined here is both anchored in practical activities through which people engage with the environment and informed by beliefs and ideas about the nonhuman world. It develops in part through human actions in the world, for instance in nurturing land through cultivation, herding animals to pasture, and a wide range of other actions meant to ensure the reproduction and vitality of the nonhuman world. These practices cannot be divorced from the inward dispositions that inform them. Ladakhi elders often express the idea that today, although people continue to pray, they "have become empty at heart," and that, similarly, many farm without putting their heart into it. This signals that what may appear at first glance to be virtuous practices may in fact be devoid of righteous intentions. Yet to be potent, moral virtues require coherence between attitude and actions.

Consideration of the social and cultural elements that inform ethical practices through which people engage the nonhuman world allows us to ponder the effects of changing political and economic circumstances on people's ethical lives. An examination of the way Ladakhis orient their lives in this context and the way they experience and resolve (or leave unresolved) certain moral dilemmas requires a perspective that looks beyond a totalizing system like religion.

What is meant here by the terms "ethics," "ethics of care," and "morality"? My understanding of an ethics of care for nonhumans follows anthropologist K. Sivaramakrishnan's notion of "ethics of nature," defined as "a set of abiding concerns and guiding principles that humans ponder, articulate, and deploy in their interactions with the non-human world" (2015:1263). However, I use the expansive term "nonhuman," rather than "nature" or "the environment," in order to include both elements of the natural landscape (such as glaciers) and nonhuman species. The notion of care refers to a mode of practical engagement informed by a sense of obligation and responsibility toward the more-than-human. For instance, when Ladakhis speak of their domesticated animals, they often emphasize the moral responsibility to care for them despite the fact that they, over the past decades, have gradually been losing their significance as an economic asset. But because domesticated animals have for generations enabled Ladakhis to ensure their subsistence in a harsh environment, many feel that caring for them is a moral responsibility. Care is therefore strongly anchored in a principle of reciprocity.

I approach ethics as a field of practices that are "socially located and culturally informed" and "undertaken with at least somewhat conscious orientation towards conceptions of what is good, proper, or virtuous" (Rogers 2009:11). In his historical and ethnographic study of ethics based on the case of a community of Old Believers in Russia, anthropologist Douglas Rogers (2009) shows how ethics is a dynamic phenomenon that can shift along historical transformations and between generations. Rogers's insights are apposite in thinking of Ladakhis' changing sense of what constitutes appropriate behavior with the nonhuman world. The transformation of Ladakhis' ethics of care is shaped by political and economic processes. Since the independence of India, the flourishing of access to employment outside agro-pastoralist occupations, whether in the civil service, the army, or related military work, means that what constitutes "the good life" now differs between generations. Inasmuch as an ethics of care develops and is sustained by practical activities through which people engage with the environment, these processes contribute to transforming that ethics by altering the way people engage with the environment.

Distinctions, or lack thereof, between ethics and morality have been the subject of debates in the literature.[7] I follow my interlocutors' framing and distinguish between ethics and morality by associating the former with the realm of practices and the latter with the realm of obligations and duties. When Ladakhis talk about the reasons for the changes in their environment—receding glaciers, eroding mountain vegetation, erratic meteorological conditions, drying of water sources—they frequently explain them as the result of people failing to fulfill their responsibilities or being careless (*tsana met kan*). Accordingly, the "moral order" stands for a collective of actions and values that pertain to a socially and culturally defined realm of obligations and duties, or what one ought to and should do. These responsibilities consist of a repertory of concrete actions in the world, such as plowing the land and herding animals or activities oriented toward divine beings, such as addressing the deities through rituals. It also consists in the cultivation of good relationships with villagers and caring for family members.

In parallel with the practical activities through which people engage with the environment, the Tibetan Buddhist cosmology, which is central to the way many Ladakhis link human attitudes and actions with the condition of the environment, also contributes to the shaping of an ethics of care. According to Tibetan Buddhism, the mountains, the soil, and water sources are all dwellings of spirits and deities that affect the fortunes of the laity; these beliefs contribute to infusing the landscape of Ladakh with morality.[8] A strong sense of reciprocity characterizes the relationships between people and the deities. For instance, in Ladakh agrarian practices have traditionally been marked by rituals that attend to the interdependence of nature, the laity, and divine beings (Bastien 1983; Dollfus 1989, 1996, 2008; Gutschow 2004). Accordingly, if the spirits and deities enable human prosperity by providing resources such as land or by creating favorable weather conditions, the laity must pay them due respect by following prescribed rituals. The presence of these deities, scholars have argued, is also intricately linked to the cosmological order that underpins the social realm in Ladakhi communities.[9] For instance, in Buddhist Ladakhis' beliefs, chthonic beings are particularly sensitive to human behavior. A lack of harmony between people and an uncaring attitude can offend these deities, who might, in the absence of a corrective ritual, decide to no longer protect humans.[10]

Although I gratefully draw from conversations I have had over the past decade with religious experts in Ladakh itself and outside, this book takes the perspective of Ladakhi farmers and herders. During our conversations, I have occasionally encountered the idea that the depletion of environmental resources results from a lack of consideration for the local deities, or from failing to perform regular propitiation rituals, but such explanations were generally postulated only by the few of my interlocutors who were well versed in religious matters. Moreover, while my interlocutors would occasionally explain adverse changes in the environment as a form of retribution by deities who are angry at contemporary Ladakhis' attitudes, most of the time they spoke of a sense of broken morality, without reference to deities. In both cases, the changing attitude was frequently described as people having become "empty at heart," something that, for Ladakhis, denotes a corrupted moral attitude and, depending on the context, may refer as well to the triumph of individualism over collective values.

In other words, my interlocutors' discourse about a changing environment is characterized neither by systematic reference to the realm of deities nor by a neat division between a Buddhist moral order, a social order, and a realm of deities. That said, the perceived root causes for these changes cannot be isolated from a feeling of decaying morality. My point here is not to divorce an ethics of care from religion. However, we should be careful not to put too much emphasis on religious doctrine in order to understand Ladakhis' moral attitude and ethical dispositions: as with any religion, adherents of Tibetan Buddhism do not go through everyday life following its precepts as one would follow a guidebook.

The notion of "fragments" postulated by anthropologist Anand Pandian (2008) is useful when thinking of the myriad forms taken by moral virtues, which seemingly have lives of their own. Studying the traditions relating to the moral virtues in South India, Pandian acknowledges the need to refer to certain traditions and discourses in order to understand actual practices and forms of moral knowledge. However, these traditions may not necessarily be coherent with their canonical foundations. An anthropological approach to ethics would suggest that traditions are "fragments," which are "often anonymous echoes from the past," that can morph into a variety of narratives and forms that pertain to different fields of everyday life. These fragments "may be understood as the very form in which the moral resources of the past survive and work as spurs to ethical conduct in the present" (Pandian 2008:467).

Fragments of traditions, whether with roots in religion or in cultural values and practice, cannot be dissociated from an ethics of care, whether it is the emerging sense of sacrifice for the land (chapter 3), the acknowledgment of a filiation with a glacier or the moral responsibility to cultivate the land despite the vagaries of agrarian work (chapter 4), the attention given to domesticated animals despite a series of predicaments (chapter 5), or the cultivation of knowledge about glaciers (chapter 6). Hence, caring for the land, for glaciers, and for animals is for many Ladakhis a moral responsibility that religious ideas about the natural world may have at some point in time contributed to shape as a moral virtue, even when they are not postulated in such terms explicitly.

AFFECT, EMBODIMENT, AND THE EVERYDAY

An attention to the practices that nurture attachment to the nonhuman world, and to how these practices are impeded by structural processes, can shed some light on "how moral and ethical worlds are formed and change as people interact with or imagine the landscapes in which they live" (Sivaramakrishnan 2015:1264). But how can an ethics of care develop and be nurtured through mundane activities—or, to make this question specific to Ladakh, within the course of everyday practices and agro-pastoralist work? Attention to the embodied practices of ethical engagement that characterize life and work in the mountains of Ladakh can shed light on the development of an ethics of care. Also, affective dispositions such as compassion and devotion, which infuse many of the moral, religious, and literary traditions of South Asia, "surface as essential means of negotiating ethical actions" (Pandian and Ali 2010:10). In other words, ethical dispositions can be seen as both informed by traditions and sustained through practical engagement with the world.

That ethical dispositions for nonhumans are shaped in the mundanity of everyday relations is something forcefully demonstrated by anthropologist Radhika Govindrajan (2018) in her account of animal intimacies in India's central Himalaya. More than the product of traditions and engagement with moral philosophy, ethical ideals in mountain villages of Uttarakhand develop through spiritual, affective, and material entanglements of humans and non-humans. Govindrajan shows how nonhumans, in their meaningful affective relationship with humans, play an active role on human subjectivity and on the shaping of their ethical dispositions, which may take various, and some-times unexpected, shapes. But intimate interspecies engagements are also affected by wider economic and political processes, with crucial implication for the ethical claims humans and animals make on one another.

Taking a cue from these reflections, I place practical interaction with the environment and ideas about the nonhuman world, informed by traditions, on equal footing in the shaping of an ethics of care.[11] In order to account for the realm of affect that develops through the embodied experience of life and work in the mountains, and therefore in the intimacy of the everyday, we may think of an ethics of care as an "embodied ethics." In emphasizing engagement with the environment, I walk in the footsteps of anthropologists who have taken phenomenological approaches to study how bodily interac-tion with the material world, in particular through labor practices, generates forms of knowledge (Ingold 2000; Marchand 2010; Pálsson 1994). But what does the embodied experience of interaction with landscapes generate to nurture ethical dispositions? While this scholarship has focused on the skills produced by the integration of the mind, the body, and the environment through labor, I am here interested in how this engagement, in meeting a world of beliefs, generates affective feelings and sentiments.

The concept of affect defines intensities that arise at the convergence of materiality and subjectivity during encounters between humans and nonhu-mans, including any elements of the material world (see Bennett 2009).[12] Affect has been described as an intensely felt, interactive bodily experience that is neither presocial nor fully social (Massumi 2002). This embodied feel-ing can be "ordinary," that is, emerging from everyday life processes (Stewart 2007). The notion of affect foregrounds mediation and thus decentralizes the immediacy that dominates contemporary theory (Mazzarella 2009). In summary, a focus on affect enables the study of the feelings that arise from the embodied experience of human-nonhuman interaction.

As it foregrounds feelings and subjectivity, affect also draws attention to the imagination with which the material world is invested.[13] Ladakh, for instance, is envisioned by many as a place of potential encounter with the divine in its various forms. The deities in question are often conceived of as transcending space, but they are also said to dwell in specific places, such as mountains or water sources. Moreover, beliefs in the reflexivity between people's attitudes and the condition of the environment contribute to infusing the environment with a moral character (P. Harvey 2000; Samuel 1993). Thus, in Ladakh, landscapes are repositories of morality that constitute and animate ethical dispositions (Basso 1996; Gold 2001).[14]

A theory of affect can thus illuminate the processes by which the sentiments that emerge through practical engagement with the environment may contribute to the development of ethical disposition. The embodied experiences of life and work in the mountains, along with ideas about nature, play a central role in shaping and nurturing an ethics of care for the nonhuman world in Ladakh. Affective dispositions are fostered for landscape and the elements that constitute them, for the land, and for domesticated animals. But the practices that constitute an ethics of care are not immune to broader structural forces. The postcolonial geopolitical and economic contexts of Ladakh have over the years led to the displacement of the agro-pastoralist economy and the disruption of the traditional community arrangements. Changing modes of engagement with the environment have crucial implications for ideas about Ladakh as a place, as exemplified in agrarian work and what results from it: ethical community relations, forms of knowledge generated by the pastoralist landscape, and the development of intimate knowledge about glaciers through herding in the mountains.

But affect is also pivotal to the production of the state in Ladakh and to the reconfiguration of the region as a border area.[15] The first Indo-Pakistani war in 1948 marked the entry of Ladakh into Indian nationhood and its new identification as a border region. The role played by Ladakhis during that war led to sympathy, on the part of the state, for this local population, a feeling that continues to infuse the defense of the territory. Moreover, through affective labor—i.e., work that aims to produce or alter one's emotional experience (Hardt 1999)—Ladakhis have come over the years to internalize their role in the military production of the state. The presence of the army calls on collective sentiments of sacrifice and infiltrates the intimacy of family; yet it is necessitated by the omnipresent threat of border incursions by Pakistan or China, and so has become by and large an accepted force in the region.

WAR AS A MARKER OF SOCIAL TIME: A CHANGING MORAL ORDER

The perceived changing moral order in Ladakh today cannot be dissociated from the historical and political processes that have transformed the region into a border area. India's independence in 1947 was marked by the partition of former colonial British dominions into the two sovereign states of India and Pakistan; the division took place along religious lines, the Muslim-majority territories being incorporated into Pakistan, while the regions comprising a Hindu majority were integrated into independent India. At the time of independence, Jammu and Kashmir was a princely state made up of a conglomeration of small independent kingdoms, among them Ladakh. Ruled by the Hindu Maharaja Hari Singh, Jammu and Kashmir at the time was predominantly Muslim, with a Buddhist minority confined largely to the eastern part of the state. During the Partition, over five hundred princely states, among them Jammu and Kashmir, had no option but to join one of the two emerging nations. The Partition resulted in widespread, brutal communal violence—riots, killings, rape, arson, and displacement of populations—that left deep, traumatic wounds on the region for decades afterward.

Ladakh remained largely untouched by all these developments until the spring of 1948, when news began to spread throughout the villages of the region that armed men were roaming the highlands, pillaging and looting along the way. Pathans and Gilgit Scouts, forces allied with Pakistan, were marching toward Leh, intending to take control of the capital. The inhabitants of Sham rallied to help the Indian troops defend the territory, thereby playing a central and largely unheralded role in Ladakhi and postcolonial Indian history. When the Pakistani raiders invaded, Ladakhi men and women guided Indian troops through the unfamiliar terrain's mountain paths, their knowledge of the place proving crucial to repelling the incursion. As well, Shammas—the inhabitants of Sham—employed various tactics that hindered the invasion, including guiding the raiders into ambushes and spying for the Indian troops.

The first Indo-Pakistani war profoundly marked a generation of Ladakhis in Sham. Many of my interlocutors lived through those traumatic events: some as members of the militias allied with Indian troops fighting against the raiders, others as civilians living in occupied villages, yet others as captives of the invading forces. And although some adopted a neutral position, choosing not to engage actively in the conflict, none were spared from the deep wounds that the war inflicted on the population. Today, many Ladakhi elders still struggle with painful memories—some of witnessing death, some of helping to perpetrate it. The shock of the war, and the violence it entailed, also shattered the local cosmology. It is therefore seen as being invested with potential karmic retribution, with current implications for the state of the environment. For some, it also sowed the seeds of a changing moral order, bringing a change in people's attitude.

My original goal in this study was to analyze perceptions of environmental changes over time, and for this reason I targeted elderly Ladakhis among my interlocutors. But I realized early on in my fieldwork that the experience of the war was a looming presence in our conversations. I often started discussions asking about the environmental changes my interlocutors had seen during their lifetime and was surprised at how many began their lengthy answers with "One day the Pakistanis came. . . ." For others, the events of the war did not emerge until later in the conversation, often following observations about receding glaciers, depleted water sources, or loss of vegetation in the mountains—yet emerge they almost always did. It soon became clear that, for many, the war and environmental change were correlated.

That link results from the way Buddhist Ladakhis' worldview shapes their interpretation of the local history. Ladakhis are acutely aware of the many political, economic, and social changes that have affected Ladakh in recent decades and tend to frame them within a rich historical context. This appreciation for the genealogy of change is never separated from Buddhist ideas about causality, which hold that the seeds of current and future events may have been sown in the past, whether recent or distant. In this way, elders often characterized the war as *tsokpo* (*btsog po*), a Ladahki term related to notions of "dirtiness" or "badness" but also carrying connotations of moral wrong-doing.

Many elders of Sham see the current depletion of the environment as "the bad luck of Ladakhis," a form of karmic retribution for the collective moral wrongs of the war. Ideas about karmic retribution are, as anthropologist Robert Desjarlais (2003:26–27) puts it, "as basic and commonsensical to Buddhist peoples as the law of gravity is to others, in which, quite simply, any moral act, good or bad, brings about a correspondingly positive or negative result, either in this or in a future lifetime." The war presented many Ladakhis with a profound dilemma. The choice to fight alongside the Indian army to defend the land meant engaging in the moral wrong of war; yet if remaining neutral might enable one to avoid bad karma, it also meant resisting the moral imperative of defending the land. Thus, for most Ladakhis, there was no one "right" decision.

Pivotal episodes in history, such as the Partition of India, are critical events that overturn an existing social order, annihilate previous modes of thinking, and create new ways of being in the world (Das 1995:6). Such events, in constituting points of reference for collective and individual suffering, are markers of social time. For my interlocutors, the violence of the war split the past from the present. By and large, the Shammas who lived through the war see this new present as defined by negative human action and resultant negative consequences. But elders' perceptions must also be considered in light of the fact that the war marked the onset of state formation in a region that had remained largely outside the scope of the bureaucratic state during the colonial era. The production of the state in Ladakh after the independence of India has been oriented toward the reconfiguration of the region into a border area, a process accelerated by and consolidated during the Sino-Indian War of 1961–62 and subsequent wars with Pakistan in 1965, 1971, and 1999. The militarization of Ladakh, together with the expansion of the bureaucratic apparatus, has generated access to employment beyond the traditional agro-pastoralist activities, generally outside the villages, opening up new possibilities for Ladakhis to pursue individual aspirations. The entry of Ladakh into Indian nationhood thus marked the onset of demographic changes that have reconfigured village life, with profound social ramifications as a generation of Ladakhis today build their life on laboring for the production of the state.

For many elders of Sham, the war heralded the onset of a new era, and as a critical event it remains central to their perceptions of the transformation— moral and environmental—currently taking place in Ladakh. In chapter 2, I speak of a "crack" in the landscape to refer to the discontinuity of social time experienced by a generation of Ladakhis. Elders often use the language of morality to frame this discontinuity. Many interpret the crumbling of village life as a sign of a changing moral order. The observance of community arrangements has traditionally been central to defining personal virtue in Ladakh, largely because such arrangements are seen as ensuring the reproduction and vitality of the environment and, therefore, the community itself. But observation of these arrangements is made difficult by new demographic realities. Ladakhi elders frame this as new social time: "Today is a different time," "Today is not a happy time," "Bad times have come" are frequent observations. There is a clear sense among elderly Shammas that the entry into Indian nationhood, as marked by the war, brought a radical transformation of village life and marked a new epoch, or a distinct phase of a community's lived experience of social and historical time that carries an affective and emotional charge.

As a form of social time, an epoch is deeply discursive, evoked to contrast the past with the present. In the context of his research in the south of France, anthropologist Matt Hodges explains that informants posited a previous epoch as the "golden age" of communal village life, "a past of organic community and enduring time," which they constantly "contrasted with a modern epoch of disenchantment and erratic time—articulated in local terms" (Hodges 2010:124). Ladakhi elders, too, contrast the past epoch with the present one by emphasizing a temporal and affective discontinuity, which they associate with a changing moral order.

The new epoch brought the growing physical and emotional isolation of elders in Ladakhi villages. While Ladakhis are unanimous about the abuses they suffered during the pre-Independence autocratic rule of the Dogra of Kashmir, there is an apparent paradox in that they often associate positive emotions with that past in contrast with the present, characterizing the previous epoch as a "happy time" that differs from the "bad time" of today.[16] But the sense of a lost past shared by many elders also interweaves with their individual experience of isolation in villages. Ladakhi elders are well aware that they have become a burden in a society in which the traditional family and community safety nets have been weakened by changes linked to far-reaching economic reconfiguration. Their repeated references to the present as being stripped of happiness are evocative of this reality.

THE MORAL NATURE OF CLIMATE CHANGE

In the past decade, climate change has emerged as one of the most critical issues in both natural and social sciences. In particular, climate change adaptation, or how to lower the risks posed by its consequences, now infuses discussions in different circles. In his study conducted in the Peruvian Andes, anthropologist Mattias Borg Rasmussen (2015) turns this question around, asking instead how climate change adapts to human lives and encounters local social, political, and economic worlds. At the local level, although people may not be actively concerned with climate change as a phenomenon, it may nonetheless exacerbate forms of stress already experienced by a community. In this, Rasmussen walks in the footsteps of anthropologists who have argued that discussions about climate change should not be the sole purview of climate science, whose normative discourse leaves little space for the rich complexity of human life in understanding the issue.

A comprehensive understanding of climate change, and therefore of climate change adaptation, requires an understanding of how its impacts infiltrate societies, their culture, and their institutions (Barnes et al. 2013; Castree et al. 2014; Crate and Nuttal 2009; Orlove 2009). Climate change, when experienced at the local level, is about how different societies encounter the state, how demographic change transforms community work arrangements, how changing water levels adds to other predicaments already experienced by farmers, and so forth.

That the spheres of climate and society are deeply enmeshed supports discussions on the epistemology of nature and culture, where, as has been forcefully argued, discursive treatment as exclusive categories creates boundaries, with crucial implications for life experience.[17] Nature is a process specific to a particular place and time, the product of asymmetries and power relations, of imaginations about nature, and of concrete material and geographical configurations (Latour 2004). In this book, nature is simultaneously social, cultural, and political. The "new" glacier introduced at the opening of this book, which was produced through the act of care when *abi* Lobsang was a young girl, is the same glacier that now recedes without the intervention of villagers, the subject of changing treatment and attitude. The politics of nature deployed by the Indian state since Independence is such that people's engagement with the mountains, and thus with the glaciers of their village, has dramatically lessened. Structural processes not only transform nature, as political ecologists have long demonstrated, but also transform how people interact with and perceive nature.

The nature studied here is also profoundly moral. As it transforms the natural landscape, climate change morphs into a temporality, a new era marked by the creation of the India state. But the temporality introduced by the production of the Indian state also meets climate change, various manifestations of which are interpreted as retributions for people's lack of care and moral shortcomings. In being defined by a range of anxious feelings, from the experience of isolation in the face of a growing individualism, to worries about the future, to disoriented social and cultural references amid life in the ruins left by an economy that preys greedily on both humans and natural resources, the epoch thus postulated by my interlocutors has in fact much commonality with the Anthropocene, or "a geologic epoch in which humans have become the major force determining the continuing livability of the earth" (Tsing et al. 2017:G1). Here, humanity, in becoming a geological force, again troubles the neat division between the realms of nature and culture (Chakrabarty 2009). In their mourning of a loss of care for nonhumans, many of my interlocutors in Ladakh are in accord with anthropologists, philosophers, and other scholars who, in the environmental crisis of the Anthropocene, are calling for the extension of care beyond humans.[18]

If, for Ladakhis, nature is moral, it is simultaneously considered objectively. The secular and the sacred, far from being incompatible, intermingle in Ladakhis' interpretation of the world.[19] Ladakhi farmers rationalize changes in the environment along the same lines as environmental scientists—linking glacier recession, for instance, to warmer temperatures and decreased snowfall, as does *abi* Lobsang. But in their view, explanations that describe exclusively physical causes and effects can only partly account for environmental changes. In the Tibetan Buddhist perspective, nature is more than its physical material manifestations. For many Ladakhis, explanations for environmental change must also include intangible, existential root causes, and these are often framed in the language of morality. This is why *abi* Lobsang was so perplexed by the snow leopard's attack on her animals: she was living her life as a dutiful person and, therefore, could not understand why she had suffered such a retribution.[20]

A PARCHED PLACE PERCHED IN THE MOUNTAINS

Ladakh is a high-altitude trans-Himalayan region in the northern Indian state of Jammu and Kashmir and was once a feeder of the Great Silk Road. It is flanked by the two highest mountain ranges in the world, the Himalayas to the south and the Karakoram to the north. The overall emptiness of Ladakh's terrain is striking, especially when approaching by plane. A bird's-eye view reveals a collection of scattered settlements, while the valleys, tucked in between mountains, occupy only a tiny proportion of space in the vastness of a territory made up of desert steppes, steep mountains, and ice summits. In this vast region, people have settled sparsely, in narrow valleys that form small oases in the cold desert. Ladakh comprises two districts, Leh, which has a predominantly Buddhist population, and Kargil, which has a predominantly Muslim population (map I.1).[21] Leh District, a territory of roughly 45,000 square kilometers, shares borders with both China and Pakistan;[22] it includes Sham, the area to the west of Leh, which spreads from Nyemo to Khalatse (map I.2).

Ladakh is located in the rain shadow of the Himalayas, which prevents the entry of the monsoon, leaving the region with scant annual rainfall of less than 100 millimeters in the upper Indus valley (Dame and Nüsser 2011:181). Agricultural production is based almost entirely on irrigation. The great

Map I.1   Jammu and Kashmir State

majority of fields are supplied by glacial meltwater and accumulated snow.[23] It is no exaggeration to say that glaciers are the lifeblood of Ladakh, a metaphor that takes on a striking poignancy in aerial photographs that reveal the morphology of villages: they seem to have poured out of the mountains (figure I.2). Land elevation in the region ranges from 2,700 to over 6,000 meters, with a mean altitude of about 3,000 meters. The physical geography of Ladakh is characterized by two distinct types of terrain: eastern Ladakh by the high-altitude plateau of Changthang (3,500–4,600 meters), its western and central parts by mountain ranges punctuated by valleys.

Ladakh experiences high seasonal temperature variation: summer days can be very hot, with temperatures reaching 35 degrees Celsius, while winter days are quite cold, dropping to –50 degrees Celsius in the coldest parts of the region, though the mean minimum temperature in Leh in January is approximately –15 degrees Celsius. Ladakh has four distinct seasons, but more often than not people speak in terms of two seasons, winter and summer, or perhaps more accurately, winter and nonwinter.[24]

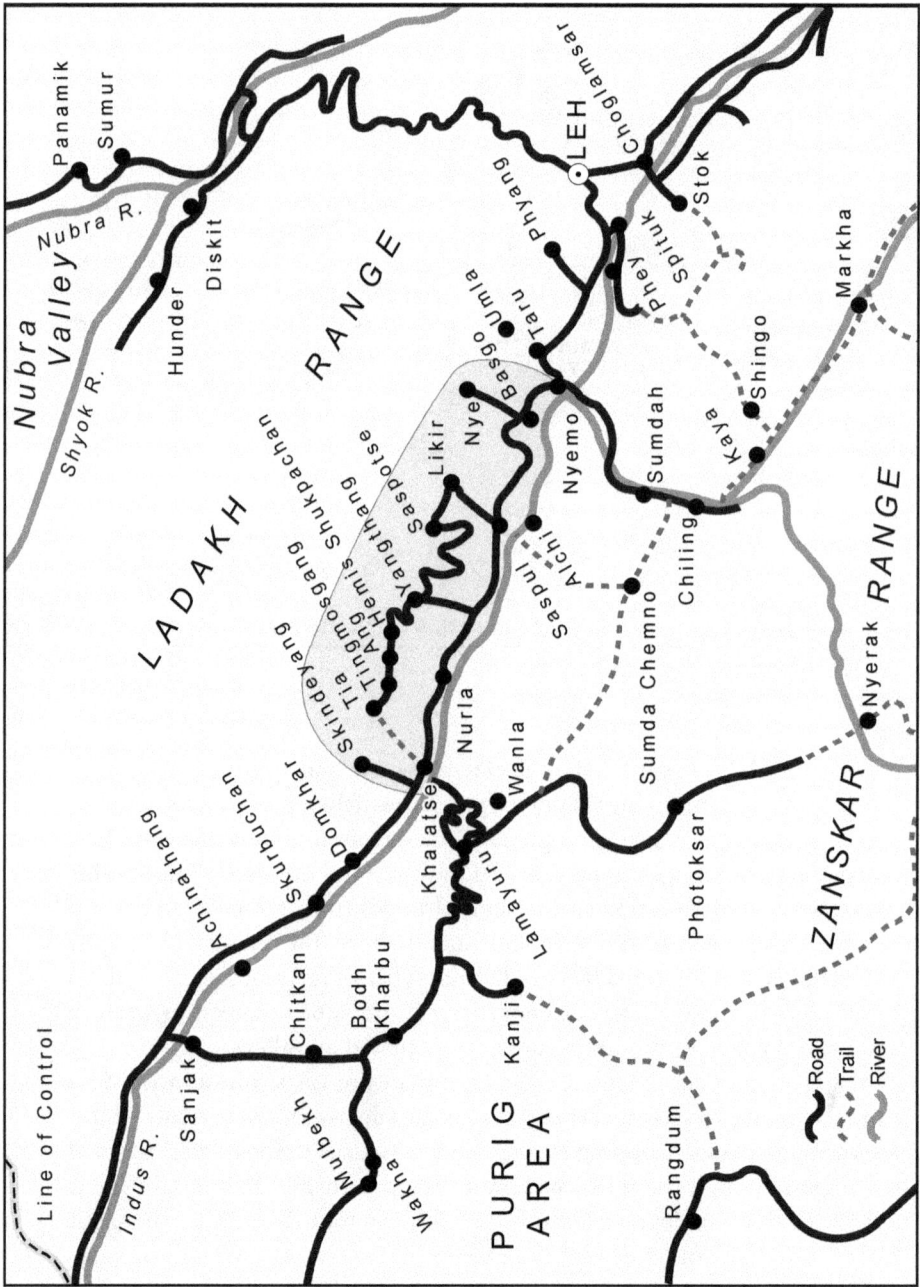

Map I.2 The Sham area of Ladakh

I.2   Aerial view of a village in Ladakh

THE METHOD AND THE JOURNEY

Upon arrival in Leh, I head, along with Namgyal, my research assistant and local host, straight to the office of the superintendent of police, where I have to register, as I am in India on a research visa. There we are met by an officer whose office, filled with piles of papers and devoid of electronic equipment, seems to date from another era. The man reproachfully informs me that had I came one day later I would have had to pay a late registration fine. "Sorry, sir," I explain, "I had to attend to other administrative matters in Delhi before flying to the mountains." He ignores my response and continues to leaf intently through a stack of papers, all profiles of other foreigners; judging by the pictures that identify them, some of the forms are decades old. The officer pulls out the profile of a man who, I read, was in Ladakh to do architectural conservation work in a monastery. He starts to fill out my form guided by this template. The architect and I apparently have bureaucratic common-alities. In the center of the room, the floor is marred with burn marks. In the ceiling, the chimney hole for the *bokhari*, the portable wood-burning stove that is so essential to any Ladakhi house in winter, is stuffed with rags. It is the end of September and the weather is still pleasant, so the woodstove has not yet been installed. Two weeks later, the bucket of water I use to wash my face in the morning is frozen.

The officer turns to Namgyal and asks him a succession of questions: Who are his relatives? What is his phone number? How do we know each other? Am I paying him to be my research assistant? How much? Am I really going to stay over the winter? And perhaps most surprisingly, do I like Lada-khi food? (I never quite understood why this concern was important to the police.) Anyone in Ladakh hosting a foreigner on a research visa has to go through some interrogation of this kind. The officer lectures me on the limits of my movements, which should exclude areas that are contiguous to the lines of control with both China and Pakistan, and I should remain outside any restricted zones or cantonment areas. "Yes, sir," I reply obediently.[25]

The officer leaves the room, explaining that he needs to see the senior superintendent of police. "Do you think everything is fine?" I ask Namgyal, who mumbles, "God knows," in the blasé voice I would later come to find so typical of his profound cynicism for bureaucratic procedures. I wait anx-iously. In theory, the local police cannot refuse my entry, as my research project has already been approved by the Ministry of Home Affairs of the Government of India. But in a border area, unpredictability is the norm. The man eventually comes back with my passport. "Here, everything is fine," he says with a laconicism that contrasts sharply with my stress over this crucial step in my fieldwork.

◆　◆　◆

On our way to Namgyal's home, located in a village on the southern side of Leh, the face of the city reveals itself as yet another overcrowded Indian city-to-be, with poor urban planning translating into queues of beeping cars. The sole larger agglomeration of Ladakh, Leh exemplifies much of the rural-urban tensions present in this region and throughout India. On our way, near the bus station, I can see houses perched atop bare, rocky outcrops built by settlers who have migrated from villages. With a population that grew from a mere 2,895 in 1901 to 30,870 by 2010,[26] Leh is something of a sad dust bowl, from which humans have visibly stripped the land of much of its resources. Encroachment on common land is ubiquitous. At night, people lay bricks on plots of land to claim a piece of ground with the skeleton of a house. Even the army resorts to the same tactics, so that today the city is flanked by an expanding number of military installations—sometimes occupying common lands—all in the name of border security.

We reach Namgyal's home. After briefly introducing me to his mother and uncle, Namgyal urges me to go to sleep to stave off altitude sickness. I step into the room that will be mine for a few weeks, before I eventually move down to the first floor for the winter. I can feel that my breathing has become labored from climbing the few steps, and I start to nurse a headache. From the window, I can see *dzos* (*mdzo*) in the field outside.[27] In the tree next to the window, a magpie perches proudly, its royal blue feathers contrasting with the brown hue of the house's mud bricks. The landscape before me is dominated by the magnificent Stok Kangri, one of the highest summits of Ladakh. In the background, the sounds of the city are singing out. From the adjacent living room, I can hear people speaking a language that sounds very different from the Standard Tibetan I am more familiar with. My head is spinning. I feel extremely tired. With the stress of my registration now behind me, the weight of high altitude hits me as if I had inhaled soporific gas. I sit on the bed, badly in need of rest. Just for a few minutes, I think. Questions run through my mind: What am I doing here? How will I be able to spend a year in this place? Will I make it? I put my head on the pillow. I wake up many hours later, this time in another world.

♦ ♦ ♦

My Leh host family was a bit reluctant to have a foreigner stay in their house over the winter. Not only that, but the mother, the head of the household, was going to spend the winter in Punjab, where her husband, a military man, was posted, leaving Namgyal and me to manage the household for the duration of the winter by ourselves. This clearly did not sit well with her—though she was more likely dreading her house being set on fire through neglectful use of the *bokhari* than us suffering from malnutrition.

After her departure, and as the cold started to set in, our quality of life dramatically declined. Namgyal's mother had turned off the water and emptied the pipes to prevent them from freezing, so water had to be fetched from the community tap almost daily, a one-way trip of a quarter kilometer, with the return trip under the weight of two full 15-liter water jugs, or 30 kilos. As the weather grew colder, the *bokhari* was installed in my small, 15-square-meter room-cum-kitchen. Namgyal instructed me on its use. Although I am quite familiar with using a wood stove, the fact that kerosene was used as a fire starter was at first frightening. But winter in Ladakh is not the time to doubt one's survival skills. Some two weeks later, I was liberally pouring kerosene into the *bokhari* to get it going as quickly as possible.

As winter set in, every household chore became excessively time consuming and unpleasant. Washing clothes in barely lukewarm water was, to put it bluntly, miserable, and items could hang on the clothesline outside for days before they dried. After a time, balancing time between fieldwork and maintaining cleanliness became irrelevant. When the number of days between baths started to exceed a week, I stopped counting them. But I was initially reluctant to let go of washing regularly. My room in Leh was infested with moths, and an episode of scabies contracted by sleeping in dirty blankets elsewhere had left me traumatized. Nevertheless, I spent days without seeing my own body, because of the difficulty of getting water and the agonizing pain of taking a bath in the cold of winter. Also, removing a layer of clothes meant losing precious body warmth and spending the next few hours shivering. And I was not the only one to search for warmth in winter: one day, I found a dead mouse under my pillow. In order to minimize the discomfort of the cold, I simply had to accept the discomfort of my lack of bodily cleanliness. I also got used to using an outhouse at −15 degrees Celsius.

Our method of work was strongly dictated by the degree of discomfort associated with winter. Sleeping next to the *bokhari* is not a common practice in the villages. But sleeping under about ten pounds of blankets is, and the body ache that follows in the morning is what I imagine aging fifty years overnight would bring. Thus, during winter, we segmented our stays in the villages with regular returns to Leh. In the villages, we were freezing, but in Leh we burned wood for heat without a second thought—something that became a problem when we ran out of wood in the middle of the winter. Facing this season requires careful preparation. Yet despite the discomforts, spending time in a village provided a unique perspective on Ladakhi life in winter, when everyone is free from farming activities and other summer labor. Village food—dry cheese and vegetables, tubers, and eggs, all carefully stored before winter's arrival—also was much better than what we were cooking in Leh, where we depended on what was available at the market.

Once the roads closed for winter, supplies in Leh started to run out as early as January, and the effect on our diet was quite noticeable.

By mid-January, the town had its first snowstorm of the season. As winter tires are nonexistent in India, cars climbed the hills with great difficulty, often sliding dangerously. One snowy day, a collision between two buses in Choglamsar, a few kilometers south of Leh, cost the lives of two people. Judging by local newspaper accounts, such accidents were common at this time of the year. That day, I remember going to the market to take in the ambience of the snowstorm and to replenish our small pantry, the contents of which were increasingly lacking diversity. As the winter advanced, fruits became nonexistent and vegetables were reduced to drab potatoes, onions, turnips, cabbages, and unappetizing carrots shriveled by frost. Our steady diet, therefore, consisted of plain rice and dhal, which is much faster to prepare than the traditional Ladakhi food. I quickly dubbed our diet "jail food," but Namgyal kept reminding me that this would have been luxury food back in his monastic years.

One day at the market, after having asked eight different shopkeepers for pasteurized milk in a box (unpasteurized cow's milk made me sick before), I finally hit a gold mine. Seizing the opportunity to buy extra on reserve, I asked the shopkeeper for four boxes, which she cautiously took from their hiding place under the counter. Only later did I notice the words "For Defense Services Only" printed across my precious find. A flourishing black market in army supplies was just another part of the Ladakhi winter.

One day, Namgyal's mother called to say that she was sending a gift from Punjab: a bag of vegetables! Namgyal's cousin was in the army and he was going to bring the package on his return to Ladakh by military plane. In anticipation of his arrival, our hungry imaginations dreamed up a bonanza of fresh food. It was with great excitement that Namgyal finally received the long-awaited phone call from his cousin announcing that the treasure had reached Leh. Bursting with joy, Namgyal asked a friend to drive him, since he thought it would be easier than ferrying by bus what we now envisioned as a veritable bounty of crisp vegetables. I waited impatiently. When Namgyal reached home, he carried a tiny bag containing exactly three small cauliflowers and a bunch of spinach. It turned out that the army was strictly limiting the weight allowable on its planes, so Namgyal's cousin had had to throw away some of the precious cargo before boarding. We ended up laughing about it, yet I knew Namgyal was deeply upset. And for the first time during my stay, I felt like crying, the deceived hope of a vegetable hoard having crystallized all the difficulties of fieldwork thus far.

Fortunately, we had another opportunity to see vitamins in the form of fresh food brought in from the lowlands. Namgyal's aunt was returning from Punjab, where her husband was posted, and she was going to bring us a bag of fruits and vegetables. This time, we curbed our excitement. Just as well. On his way back from his aunt's place with our treasure, Namgyal bumped into many relatives. Tradition obliges Ladakhis lucky enough to receive food from the lowlands in winter to share with "relatives"—a very extendable term. When Namgyal told me how much he had given away, I was surprised to find myself feeling greedy. "What is happening to me?" I wondered.

This was one instance among many when the conditions of fieldwork drove me to breach my own set of values. The difficulties associated with everyday life in Ladakh were compounded by feelings of isolation due to infrequent communication with people back home in Canada: with a nine-hour time difference between India and Montreal, and perpetual internet and electricity outages, calling home proved to be a logistical puzzle. The emotional strain was made even greater by some difficult moments in the field. My main interlocutors for this research project were elders, which proved to be far more emotionally challenging than I had anticipated. When coming back from conducting interviews, Namgyal and I were sometimes so saddened by what we had seen and heard that we could not talk for whole evenings. Silence was sometimes the only way to process narratives that were bringing us to reconsider our perspectives on birth, death, and how the in-between is lived. In the Tibetan conception of the life cycle, old age should be devoted to preparation for the next incarnation, through the accumulation of merit (Childs 2004:131). This is also a time, for many elders, to look back on their life and reconsider certain episodes, and try to appease past wrongs through prayer. Perhaps there is something universal in the search for redemption as death gets closer and in the sadness of looking back on one's own life with regret. These are delicate existential matters that, when confided, are profoundly moving. It was only with Namgyal's help that I was able to complete my fieldwork with the objectives I had set out to achieve, and so he is a recurring character in the pages that follow.

THE PATH

Set in the Sham area of Ladakh, this book reports the views of my main interlocutors, Ladakhi elders, on the root causes of environmental change and natural resources depletion in the region. Their testimonies illustrate the historical and contemporary events that have, in their experience, contributed (and continue to contribute) to altering the landscape of Ladakh.

This is a story about the moral and ethical implications of the reconfiguration of Ladakh into a border area. It is a story anchored in a postcolonial geopolitical order, involving territorialization, nation creation, and the enforcement of new borders, often through militarization.[28] Ladakh shares some of the most contested geopolitical borders in postcolonial South Asia. My aim is to explore the wide-ranging moral and ethical implications of this geopolitical situation, which contributes to altering the material and affective landscape of Ladakh, whether these changes are located in interactions with the state or in how Ladakhis engage with Ladakh as a place. The book therefore proceeds in a diachronic fashion. First, it examines the historical events surrounding the establishment of the Indian state in Ladakh, which contribute in turn to the affective basis of the subsequent military production of the state. It then looks at the moral and ethical implications of the ever-growing presence of the state by emphasizing the tensions and the dilemmas produced by the displacement of the traditional agro-pastoralist economy. It is here that climate change enters the story, as its manifestations in the landscape give rise to reflections about a fallible humanity.

The book's structure mirrors the cycle of the seasons, from the beginning of autumn to the end of summer. In doing so, it reflects the rhythm of social time in Ladakh, while describing changing modes of engagement with the environment. It provides insight into how an ethics of care develops, how it is nurtured, and how its preservation is compromised by broader political and economic processes. In this regard, it speaks of a reality that transcends the limits of Ladakh itself, a small community perched in the Himalayas.

The Loneliness of Winter

Continuity and Change in the High Mountains

It is really good times for the children these days, but for the old people it is very miserable and sad, because they are all alone.

—TSERING DOLMA, NYE VILLAGE

Though it was tougher then, people were happy at heart.

—TASHI YANSKIT, BASGO VILLAGE

ENTERING SHAM BY FOLLOWING THE INDUS DOWNSTREAM ABOUT 30 kilometers from Leh, the visitor is greeted by the magnificent sight of the confluence with the Zanskar River. A 2-kilometer trek along the Indus's immaculate, gray-colored sandbanks leads to Nyemo village. The water, which turns brownish in summer, when the surrounding glaciers are in their seasonal melt, slowly becomes turquoise in autumn, before partly freezing in winter. In the absence of central heating, winter can be a time of physical discomfort, but this is offset by the charms of the beautiful wintry landscape. By early January, the surrounding summits have become white and the fields are covered with snow. With this winter landscape comes a strange silence. The ambient noises are muted, as if absorbed by the white blanket. The winter cold seems to slow the pace of all activity. Time appears to stand still. The small village of Nyemo feels quasi-abandoned. Elders sitting in the warmth of the afternoon sun are just about the only souls to be seen. The villagers who have not deserted Ladakh for the winter take refuge in their kitchens, the only comfortable part of a Ladakhi house at this time of the

1.1   Harvesting potatoes by the Indus in Nyemo in October

year. Even Leh town starts to feel like a ghost town as early as November. As winter creeps in, all activities become increasingly difficult.

To understand the ramifications of these seasonal changes on Ladakhis' engagement with Ladakh as a place, we must first look at the elements of the traditional organization of life and work in the region. This requires considering the household as a site of integration for various economic activities, something made possible only through a division of labor among extended family members. Seasonal outmigration is one of several patterns of mobility adopted by Ladakhis today, which brings a wide range of experiences, often contrasting along generational lines. Along with rural depopulation, seasonal outmigration shows that the traditional Ladakhi household is no longer the fundamental site of economic integration. For many, the pursuit of aspirations often entails physical departure, sometimes partial, sometimes complete, from a unit that not so long ago was the only means of survival in this region.

The nature of the Ladakhi household as a locus of social integration, in particular for different generations, is also today seriously compromised. In both academic and popular discourses, descriptions of Ladakh often revolve around the trope of remoteness, with the region being painted as extremely isolated geographically. More than the product of a distinct geography, the experience of remoteness is often the result of historical processes, political violence, and discursive constructs (Hussain 2015; N. Mathur 2015). Moreover, imaginations about edges and remote spaces are inseparable from conceptions of centers (Harms et al. 2014:363–65). Although from the economic and political perspective of Delhi, Ladakh may have long been a remote area, the feeling of isolation experienced by Ladakhi elders today is only exacerbated by state interventions that, since the independence of India, have aimed at integrating the region with the rest of the country. These measures have ramifications in the Ladakhi household, which is losing its nature as a site of continuity and interdependence between generations. Many elders increasingly experience Ladakh in terms of isolation, as family members and neighbors move outside the villages for extended periods of time.

"Loneliness" here refers to the profound sense of isolation elders feel amid what many describe as "empty houses," as well as a feeling of having become a burden. Articulation of this feeling often gravitates around the perception that household dynamics and communal life are no longer the same, as people have seemingly become more preoccupied by their own interests than by those of their family and their community. For elders, loneliness is a product of a younger generation's social and economic aspirations, which are closely linked to the increased presence of the state. Since India's independence, Ladakh has benefited from infrastructure development and poverty reduction schemes, and has seen increased access to employment in the state apparatus, whether in the army or the civil service, which by and large implies long stays outside the villages or even relocation to Leh. But overall, because it has not proceeded at the same pace as elsewhere in India, development has generated unmet expectations, especially in the rural area, which has become increasingly conceived as an outer fringe with limited opportunities. More than any geographical disposition, it is precisely dissatisfaction with the absence of development that often leads younger generations of Ladakhis to depict Ladakh as a remote place. Many lament the lack of access to resources in the region, the poor quality of communication infrastructure, and the physical discomfort as well as lack of sources of income during winter.[1] Elderly Ladakhis, meanwhile, often emphasize how government investment in the development of the region has brought an easier life for younger generations, while they find themselves increasingly alone.

To be sure, the distinct winter climate of Ladakh imposes its own rules, which a growing number of Ladakhis are trying to avoid today. Winter is not unlike a long sleep.[2] Time stands still and a calm prevails, seemingly broken only during the festivities of Losar (*Lo gsar*), the Ladakhi New Year. Winter is a time of making the best of a difficult situation: enjoying the company of family, friends, and neighbors, talking for hours while drinking *chang* (a homemade barley beer) or butter tea. Without this dimension of social life and the rewards of social cohesion, winter can easily become a tremendously arduous experience. But this is becoming the reality for many today.

Movement outside Ladakh in winter is a relatively recent development. If perhaps not widespread, this phenomenon is significant enough to have arguably taken on an aspect of social capital, primarily because spending the winter in Ladakh signifies hardship. Among other things, winter means not having access to fresh fruits or vegetables and undergoing the sometimes intense physical discomforts of such ubiquitous Ladakhi winter chores as fetching water from a frozen stream, washing clothes by hand in frigid water, feeding the *bokhari* with wood, shivering at night despite sleeping under a heavy pile of blankets, and, worst of all in my experience, bathing in the winter cold. In addition, especially for those who do not have access to farm produce, life in winter is quite expensive.[3] For families like Namgyal's, who do not have a tree plantation, buying wood for the *bokhari* is a major expense. Indeed, we experienced it first-hand during the winter: when our supplies ran out, in late January, we needed to restock our firewood, but locating a source, finding a place to chop the logs to a size that could fit in the *bokhari*, and hiring help to carry the loads over the very uneven ground separating the road from the house all proved to be a logistical nightmare that stretched over many days and turned out to be a surprisingly costly affair.[4]

A significant group of Ladakhis who remain outside the region during winter is constituted of youths attending college or university; increasingly too, those who can simply afford to do so, especially among younger generations, spend the season in more temperate places, such as Jammu, Delhi, or Goa. In addition, the organization of livelihood around paid work, probably the most significant difference between generations in Ladakh, takes people out of the villages for many months or most of the year. While government employees, in particular teachers, are posted to various locations throughout the region, many others, such as those working within the ever-expanding government bureaucracy, have to relocate to Leh, and men employed in the army spend most of the year away from home. Overall, the population movements redefining present village life are varied and have significantly contributed to transforming village and household dynamics. The effects tend to be more evident in the harshness of winter, which can transform simple everyday activities into a challenging task.

Nyemo village is located on the bank of the Indus and thus along the same main road that in colonial times travelers, caravanners, and officials on duty took to Leh or Kargil area and beyond. One of only two roads that connect Ladakh with the rest of India, the National Highway 1D, crosses the village, having paved over what must once have been a bucolic settlement. This highway connects the town of Leh to Kargil District; the second major road links Leh to Manali, 480 kilometers away in Himachal Pradesh state. By November, both these roads are closed when snow covers the high mountain passes, and they open again only around May.[5]

When India became an independent country in 1947, the economy of Ladakh was based largely on subsistence agriculture and livestock rearing. Barter was also a crucial element of local economic relations, essential to providing households with necessary supplies. In Sham, the trade of local wool for various commodities from Kargil, Suru Valley, and Baltistan, as well as Changthang and Gertse in Tibet, was fundamental to the Ladakhi household economy. Each of these occupations, cattle rearing, agriculture, and trading, constituted a way of engaging with the environment and knowing the territory in this part of the Himalayas.

Today, as in the past, Nyemo village is an almost inevitable stop for anyone traveling through Sham toward or from Leh. In the era of trade caravans, the village was a supply point where travelers would buy various provisions for themselves and feed for their animals. And as a regular stop along the bus route, Nyemo maintains this function to some extent today. Several stores cater to travelers, most of whom stop here only long enough to have a meal and do some quick shopping. Buses and trucks come and go, their old engines leaving behind huge clouds of thick dark exhaust, which dissipates only very slowly in this high-altitude thin air. Adjacent to the village is a vast and charmless military camp.

Behind this frantic transit hub is a tranquil village filled with memories of the past. Tsewang Jorgais has lived all of his ninety-one years in Nyemo. When he was young, the village's population was a fraction of its present size, which today stands at about a thousand inhabitants, according to official census data. The official numbers, however, are only a rough approximation of the reality on the ground, as significant proportions of villagers today spend more than half the year outside their home village. Tsewang Jorgais was already a married man when India gained independence in 1947. Sitting outside by the mud brick wall next to his house, he wears a brown *koncha* (*gon cha*), the traditional Ladakhi long woolen coat; a turquoise earring; and a long beard. Peeking out from under his wool hat is a small ponytail, thinned by age. The man still wears his hair in the style of the old days: shaved at the front and long at the back.[6]

Tsewang Jorgais is old, and he feels old: "I have lived beyond what is needed," he tells me as we sip butter tea under the sun. He is tired of being trapped in a body no longer strong enough to cope with even the most mundane matters. "My eyes are so weak now, I cannot even see your face," he confides. Loss of eyesight affects both old and not so old in Ladakh, thanks to the long-term effects of exposure to intensely bright sunlight, which leads to development of cataracts that often remain untreated. Tsewang Jorgais's wife, Tsering Dolma, is only a few months younger than her husband and equally blind. "Both of us are of the same type," he observes. "Both of us survived through very difficult times, and we are waiting for death to take us." Although the years have not stolen Tsering Dolma's ability to speak, her desire to do so is not apparent. She sits silently next to her husband. The couple's attitude toward death reflects that of many elders in Ladakh. The pain of aging in this harsh environment and the consciousness of inevitably becoming a burden for others translate for many into a readiness for death, and elders like Tsewang Jorgais are often explicit about it.

The difficult living conditions elders experienced under autocratic, pre-Independence local rulers have also left scars. As an elderly lady once told me, "We saw so much hardship in our childhood that we became old really quickly." For some, aging also comes with a profound feeling of loneliness, a situation as true in Ladakh as in many places throughout the world. "I have seven children," explains Tsewang Jorgais, "and there is no one nearby, except one daughter who lives in Nyemo." The rest of his children live in other villages, and it is easy to deduce from Tsewang Jorgais's tone that they rarely visit their parents.

Tsewang Jorgais's life, in his words, has been that of a sad man burdened by perpetual debts—a recurring narrative among the elders of Ladakh who once lived under the rule of the Maharaja of Kashmir. As a result of a long history of autocratic regimes, until two generations ago Ladakhi farmers were inescapably "born in debt, lived in debt, and died in debt" (Sheikh 1999:345). The written history of Ladakh contains numerous gaps, but the available sources show that in the eighth century Ladakh was under Tibetan suzerainty (Rizvi 1996:56). The boundaries between Ladakh and Tibet were established in 1684, after the death of the Fifth Dalai Lama (Rizvi 1996:73–74). The succession of kings who continued to reign over Ladakh nonetheless had to pay a small tribute to Tibet, which maintained a (loose) suzerainty over the region. This came to an end following the capture of Ladakh by the Dogra army led by Zorawar Singh in 1834.[7] The takeover was sanctioned by the British colonial regime, which saw Ladakh as a "natural border" for India.[8] In 1846, the treaty of Amritsar officially transferred Ladakh to the Dogra ruler, albeit under British suzerainty.[9]

Under the Dogra rule, which ended following the independence of India, Ladakhis were taxed unscrupulously.[10] The tax load had to be paid partly in cash (*bhap*) and partly in food grain and firewood (*jines*); these goods were stored in state-run supply depots and sold to trade caravans or distributed as rations to government employees (Rizvi 1996:87). While forced labor (*begar*) already existed under the rule of the Ladakhi kings, it became a systematized feature of life under the Dogra rule. Peasants were forced to fulfill *begar* obligations, including working as porters, herders, and laborers for the state, the monastic estate, and for official and nonofficial travelers.[11] *Begar* obligations, as well as the collection of taxes for the *tehsildar*,[12] were supervised by local officials, who often took advantage of the system to enrich themselves at the expense of peasants. Intimidation, mistreatment, and violence perpetrated by officials was widespread. Far from being elusive, as can be the structural violence of bureaucracy that frequently characterizes the modern state, the violence of the Dogra state was not only physically real, but also all the more inescapable because it was often perpetrated by one's neighbors.

Begar obligations left lasting marks on Ladakhis' memories. Many elders remember with incredulity how when the British traveled in the region, Ladakhis had to carry individual palanquins bearing not only them and their spouses, but also their dogs. The work's physical strain was exacerbated by abuse on the part of government officials, who would beat the Ladakhi porters with rods if a palanquin did not stay level on the steep mountain paths. Overall, the Dogra are remembered locally as greedy for both territory and resources, as well as for their often-brutal imposition of forced labor (Aggarwal 2004:35; Bhan 2014:29; Bray 2009:49). Elders in Ladakh today are unequivocal about the hardships of this period, which they roundly describe in terms of exploitation. "How miserable it was," remembers Tsewang Jorgais. "We would come back from *begar* work at night in winter and it was so cold that our beards and mustaches would be stuck with ice."

The taxation system generated a vicious cycle of debt, in which peasants were forced to borrow from richer families in order to meet tax levies, at interest rates as high as 25 percent, and it was common practice for moneylenders to treat debtors as bonded laborers. The peasantry's living conditions were dire, with accounts of hunger common among Ladakhi elders. The leanest time of the year was during the summer months, when stocks were exhausted and the new crops had not yet matured. As a local proverb goes:

CHAPTER 1

"July is the time when flies collect around your mouth. It is the time to say, 'What shall I eat' and not 'What shall I wear'" (Sheikh 1999:346). Resentment is still palpable among elders who experienced the hardships of the period. In the words of Tsewang Jorgais, "As we did not know how to read and write, we Ladakhis were used and made fools of by the Khachul [Kashmir] government."[13]

To pay his due, Tsewang Jorgais had to borrow from the Nangso house,[14] a well-off family of Nyemo, and pay back in kind with labor. He was not the only one: people would come from villages kilometers around to take loans from the Nangso house in the form of grain. As the old man makes clear in an emotional reflection, once one's life was caught in the vicious cycle of debts, freedom was difficult to regain: "The rich [moneylenders] would take us to clean their toilets and dig their stables almost every day. The people who would give us loans would put us to work all the time. In those days, we were beaten and had no time for ourselves. The rich would not pay us even an *anna*[15] for working all day, except for a meal once in a while. We had to work for the rich all spring and autumn."

Governance from afar characterized the Dogra rule. The *tehsildar* was permanently based in Skardu and came to Leh only for the summer months. But the authority of the Dogra ruler nonetheless felt real to the villagers, who knew an official visit meant trouble for them. Tsewang Jorgais recalls how when the news arrived that high revenue officials had crossed the Zojila pass and reached Kargil (about 200 km away), every villager was put to work to prepare for his arrival, while those who refused to obey would be thoroughly beaten with sticks. Peasants had to fill the depot of Nyemo with dairy products, wood, and sacks of grain. Tsewang Jorgais's job was to find firewood, the precious combustible indispensable to survive the winter; to do so he had to walk all the way to the gorge (*rong*) of Chilling and carry stacks of wood on his back over a distance of more than 20 kilometers. The trees of Chilling had large trunks and thick roots, since, as Tsewang Jorgais notes with desolation, they "had been there since the time of the *abi meme*."[16] Decades of ecological practices were overturned in order to satisfy the oppressive levies commanded by the Dogra ruler and his officials.

Elders' discourse suggests that scarcity made all resources invaluable. Many spoke as if each grain, each piece of firewood, each edible plant was indispensable to the survival of the family unit. In the testimonies of those who lived through this era, there is a striking contrast between the time when authorities unscrupulously divested already destitute people of precious resources and the postcolonial "era of plenty," when "gifts from the great government," such as subsidized food rations, changed people's lives.

The contrast between the two epochs is also expressed in terms of social resources, which from the perspective of many elders have followed a reverse path, shifting from being abundant to scarce, so that a feeling of isolation prevails today. The moral imperative of taking care of one another is seen as being increasingly lost today.

Ladakhis did not remain impassive to the accession of the Dogra regime, which was followed by a series of revolts. The insurgents' violent repression eventually ensured the Dogra's complete control over the territory (Jina 1994:24), which was consolidated by maintaining the population in a state of crushing poverty. In accounts dating from the colonial era, British brigadier Louis Gompertz (1928) and Moravian missionary Walter Asboe (1947) attributed Ladakhis' lack of resistance to the Dogra regime to the Buddhist belief in nonviolence and a "fatalistic view of life." However, reading the situation against the backdrop of prevailing abuse and abject poverty suggests that both intimidation and the daily effort required to ensure basic subsistence may have been important factors in villagers' resignation. Recollections by elders of the past make it clear that saving sufficient resources to get through winter was then an ever-present concern. More than fighting against the Dogra, peasants' preoccupations revolved around how much grain would be confiscated, and whether the food stored away was going to be enough to get through winter.[17]

In order to be at the service of *sahibs*[18] and government employees arriving from Kashmir, every household in Nyemo and surrounding villages had to send one man to stay at the depot for a month on a rotational basis. Conditions at the depot were arduous: Tsewang Jorgais remembers how the men could barely keep themselves warm at night with the officials' horses' saddle rugs. It was during one of these rounds of depot duty, Tsewang Jorgais recalls, that a fellow villager, exasperated by fatigue, flung the luggage he had been forced to carry against a rock. He was heavily caned for his insubordination. Decades later, the beatings and humiliation that Ladakhis endured are still open wounds to their dignity. As Tsewang Jorgais confides, tears streaking the wrinkles on his cheeks, "Had it been just this [forced labor], and not getting beatings, it would have been tolerable. But we had no choice." Tsewang Jorgais wipes his eyes while his wife murmurs, "*Om mani ped me hum*," a mantra that resonates throughout the mountains of Ladakh.

Recalling the abuse of the old days, Tsewang Jorgais makes a clear distinction between pre- and postindependence India: "All this was before the great Hindustan," he notes, in a reference to India frequently encountered among those of his generation. "After, with Pandit Nehru," he continues, "things changed, and today the times are good and comfortable for the younger generations." Ladakhi elders share a distinct feeling that the Indian state extirpated them from misery. State development initiatives, including road building and other infrastructure projects, poverty reduction schemes such as food subsidies, and new employment opportunities in the government bureaucracy and with the army are frequently referred to by a generation of Ladakhis as gifts from the "great" and "generous" government.

Because he lives in a village by the main road that follows the Indus, Tsewang Jorgais long provided food and shelter to traveling Ladakhi traders. Villagers from Chilling area in the southern part of Ladakh, reputed for their skills at bronze craftwork, often made their way toward Leh in winter in order to supplement their meager incomes by selling their wares. Despite their own destitution, Tsewang Jorgais's family, as well as others in Nyemo, would take in these travelers, whom Tsewang Jorgais remembers as "so poor," and who dared to venture out into the harsh winter. Recollecting these memories as we are sitting under the midday sun, Tsewang Jorgais pauses to spin his prayer wheel. The creaky sound of the object animates the silence. His wife, Tsering Dolma, counts the beads of her rosary, praying silently beside him. "I did a lot in my life," says Tsewang Jorgais, his voice trembling from old age. "And then we went under Hindustan, and the road got built," he continues. After a pensive silence, he adds, "Since then these people never come here, as they took the bus to Leh from Nyemo without stopping overnight." Paradoxically, therefore, while development has connected Ladakhi villages to Leh, to the rest of the state, and eventually to the rest of India, for some it has also brought a sense of loneliness in its sway.

CYCLES OF SEASONS, CYCLES OF LIFE

In Ladakh, the physical features of the rocky mountain summits act as sun markers that indicate changes in the seasons. Villagers use these markers to determine when to perform different seasonal farming practices. If the summits have no defining features, villagers will climb up and pile stones to build these sun markers (*nyi tho*).[19] In Nyemo, one such natural marker is formed by three sharp peaks in the mountains that bound the Indus to the south of the village. This geographical feature is named Betso Shi Sa (literally, "place of the dead calf").[20] According to elders, back when winters were harsher, colder, and brought significantly more snow than today, many calves would die before spring came. When the sun begins to set at the three peaks, calves are said to no longer die of cold. It is also said that from that point onward, as the weather gets warmer, the lives of old people become longer. Winter was so harsh in the past that it constituted a critical time for vulnerable humans and animals alike. Children were also particularly at risk: accounts of child loss, still painful several decades later, mark the winter memories of many elders.

Despite the far-reaching changes of the recent decades, life in Ladakh is still conditioned by the enduring cycle of the seasons. Preparation for winter, when resources become scarce as the land slowly freezes, continues to necessitate dedicated attention, although access to market commodities has helped to reduce this task. At the end of the farming season, once the threshing and winnowing of harvested grain is completed, people are busy grinding grain into flour, fixing houses and irrigation channels, collecting various edible plants, stocking grass for the cattle, gathering firewood and dung in the mountains, drying apricots, vegetables, and cheese in the sun, and storing tubers inside the ground.

Regardless of the development Ladakh has seen in the past decades, winter continues to be harsh, and in the context of old age and isolation, winter conditions could hardly be more austere. Aging is fraught with difficulties in a place where the infrastructure remains underdeveloped and where pension and institutional safety nets are still minimal, at least for a certain generation of elders. The case of my host family in Nyemo is a telling example. The father, Morup, and the mother, Diskit, both in their early seventies, clearly lead a meager existence. Two months before my arrival, their daughter-in-law, Angmo, had moved in. Her husband is serving in the army and is posted to a far-off town in Ladakh. She has a two-year-old daughter and a newborn baby. Although she normally lives in Leh, where she is a government worker, she is spending her maternity leave with her in-laws. This way, she can ease the burden of work for the elderly couple, for whom everything from household chores to the care of sheep and goats is all the more exhausting in winter. Fetching water in winter itself is no easy task, especially since hand pumps freeze up. The local stream is a downhill walk about 500 meters from the family's house. Since the stream is frozen, a small canister has to be lowered through a hole in the ice to bring water up as many times as is necessary to fill big jugs, which then need to be hauled back uphill to the house. This, and cutting enough wood to cook and keep the kitchen warm, is truly grueling work.

Fifteen years ago, Morup suffered a cerebrovascular accident that left him partially paralyzed. He can no longer walk, and he lost much of the use of his hands, which were left twisted by the episode. The couple's house is in no way adapted to his condition. Morup has to crawl from one room to the other, which has become increasingly tiring with age. The old couple lives in difficult conditions. Their children are trying to support them as best as they can, but what they really need is for someone to live with them permanently to help with everyday needs. Yet, as with Tsewang Jorgais and Tsering Dolma, all their children are far from Nyemo: their daughters are married and live in distant villages, and all their sons are in the army. The traditional Ladakhi household structure used to provide social protection for aging family members as they became more economically unproductive. But this security net is today rendered increasingly fragile by an economy that drives people outside the villages for extended periods of time. Traditionally, too, elders continue to play a vital role in the life of the household by, for instance, transmitting knowledge or taking care of the youngest family members. These roles provide aging people with a sense of purpose, one that is however today often lost in households with no generational continuity. Having developed a friendship with Angmo's husband during my stay in Ladakh, I could see how the situation was for him a great source of preoccupation. He felt it was his duty to take care of his aging parents, but doing so would require that he quit his job, a nearly impossible move for this young father of two.

Morup and Diskit's house is quite modest. Its five rooms, small in size, all their walls painted the color of ochre, are sparsely furnished. The exception is the kitchen, whose every inch is occupied. Crammed in this room are a traditional Ladakhi bench, a small table, a *bokhari*, and a gray metal shelf supporting a television set, the only element of the house that suggests that time has not stood still since the couple's youth. Electricity has been installed haphazardly: wires hang everywhere, interlinked with connections that would make an electrician shudder. The house is perpetually, bone-chillingly cold. Most Ladakhi dwellings are built so as to receive maximum sunlight. That's how the couple's first house had been built, but a flood washed it away in 1970.[21] When they rebuilt, their options were limited, as the result makes quite tangible. One night, I slept in my winter jacket, a robust piece of apparel meant to resist temperatures as low as −50 degrees Celsius, yet I still felt cold. The discomfort, however, is pleasantly offset by the warmth of the house's occupants. Mornings bring the strong herbal scent of juniper from Morup's incense burner as he performs *sang* (*bsangs*), a daily ritual conducted in many Ladakhi households, during which the fragrant smoke of juniper twigs is offered to the local gods and spirits, who are also invoked through recitations. Intended to elicit "positive influences for the good fortune of the participants," *sang* is an example of how Ladakhi local religious practice reflects the everyday pragmatic dimensions of Buddhism (Samuel 1993:183).

Winter is indeed a deeply embodied experience, with a distinct ambiance and a hypnotic rhythm. The cold forces people to take shelter in search of warmth, curtailing most activities to a minimum. When a day is sunny and warm enough, people sit comfortably in the sun, spinning wool and mending clothes. Otherwise, a lot of time is spent in the kitchen, next to the fireplace, where people converse for hours. The harsh conditions leave lasting marks, too: cold-related rheumatism is a common health problem, one from which Diskit badly suffered.

True to a Ladakh winter, where the kitchen becomes the center of each family's universe, our days with Morup and Diskit are spent in the confines of a room no larger than 9 square meters, its walls covered with soot. The cramped, dark room is kept warm by a *bokhari* continually fed with cow dung and wood. We pass our evenings chatting and watching Indian soap operas and Bollywood movies on television at the rhythm of the power shortages. Because processed foods are generally absent from the Ladakhi diet, at least in villages, cooking everything from scratch means that a great deal of time is dedicated to preparing meals. Ladakhi cuisine is relatively straightforward and relies on a limited selection of ingredients, but Angmo's cooking has for me the therapeutic effect of comfort food, a particularly welcome change from the bland diet we had become used to in Leh since the onset of winter.

It is from this kitchen that, one night, the extent of Morup and Diskit's vulnerability to the vagaries of everyday life is brought to my eyes. While we are sitting around the *bokhari*, we suddenly hear loud steps on the roof, directly above us. "What is that?" asks Namgyal worriedly. "A dog has probably managed to climb there," Diskit calmly answers. Morup nods, approving her suggestion. A while later, I go to use the toilet on the upper floor, which opens onto a part of the roof. When I reach the bathroom, I see that a *dzo* is supping on hay that is stored on the roof. The scene does not appear overly strange to me: as the house is built on a slope, it seems plausible that the *dzo* had simply stepped onto the roof. The animal will merely make its way down, I am thinking. In any case, I figure, *dzo* are to be found just about everywhere, sometimes in the most unexpected places.

I go back to the kitchen and, without mentioning my encounter, sit down to enjoy yet another lovely meal. When the sound of steps returns after some time, I casually inform our hosts that a *dzo* is on the roof, and not a dog. From the sudden change in everybody's facial expression, I realize that this is not as normal as I had imagined. I immediately feel like a fool for not alerting our hosts to the situation. As it turns out, the *dzo* had come in through the back door of the house and climbed the stairs to reach the roof, where the hay is stored. And apparently this is not the first time the animal performed its burglary tricks to satisfy his stomach. Everybody rushes to the roof, even Morup, who has to crawl his way up the stairs, leaving me to keep an eye on Angmo's young daughters. Angmo calls Stanzin, their neighbor, for help, and he proves to be the hero of the situation: after half an hour of everyone pushing and pulling, he takes the beast by the horns and, with great effort, pulls him down the stairs with ropes.

• • •

That night, in our small room, Namgyal and I stare for a long while at the ceiling in silence. Time and again, small pieces of dry mud, used to pack the thatched ceiling, fall on us while Diskit walks on the roof, checking her animals one last time before going to bed. It is cold and the blankets are heavy. My mind is busy reflecting on the elder couple's autonomy: What would Morup and Diskit have done if they had been alone tonight? If Stanzin had not been at home? They would have managed, somehow, as they have to when they are on their own, but it surely would have been far more complicated. When considering Morup and Diskit's quotidian situation, one feels a sense of powerlessness and of abandonment. The silence of the room is filled by the sound of the crackling fire in the *bokhari*. I feel awful for being there and using the family's wood, such a precious resource in this place, one so strongly infused with moral values that my financial compensation has no bearing. "Winter is so sad," says Namgyal, breaking the silence. "All these old people left alone—I never realized before how winter is for them," he confides. "Now I cannot stop thinking about my own grandparents, all the way back in their village, alone," he says, obviously guilt-ridden.

THE BIG SPLIT: AFTER POLYANDRY

"This is my advice to you. If you want to understand changes in the environment in Ladakh today, think about how everything has become split." Being the *onpo* (*dbon po*), or the local astrologer of Nyemo, a craft he learned from his father, Tsering Tundup, a man in his early seventies, has always played a crucial role in the life of the families of Nyemo, providing people with the most auspicious date on which to perform symbolically significant rituals. But according to Tsering Tundup, family life is no longer the same, and the selfless commitments people once had toward one another are no longer fulfilled. "Everything has become split," he repeats. "The families are split . . . people are divided, not united."

Tsering Tundup's perspective on adverse environmental change as an outcome of people's growing individualism and increasing disengagement toward close kin is widespread among Ladakhis of his generation. This sense of a split, moreover, has objective ramifications in the traditional Ladakhi household structure and the organization of work in Sham, where the household used to be akin to a cooperative unit. Within the extended family, one brother would normally be in charge of supervising work on the land, while another would be in charge of pastoralist activities. Before the closing of the border with Tibet and Pakistan, many families of Sham also had one member who would ply trade routes for a significant part of the year. This diversity of occupations was a way for villagers to draw what benefits they could from the region, where arid conditions always limited the possibilities for agrarian development.

This diversity of occupations within families was made possible by the prevalence of extended family units, which in Ladakh were based in polyandrous unions. Functionalist explanations of polyandry have been challenged, yet as Bauer (2004:22–23) aptly notes, the emic perspective on polyandry nevertheless clearly favors functionalist arguments.[22] Time and again, Ladakhis explain the rationale behind this custom by referring to its imposition by a king of Ladakh as a measure of prevention, considering the limited agrarian resources in the region.[23] Polyandry also meant that traditionally in Ladakh, men's life paths were closely defined by sibling position, while a woman would, as a general rule, become a wife within a polyandrous marriage. Land estates and households were passed in their entirety from one generation to the next. The eldest son inherited the family estate by right of primogeniture, while the options open to his brothers were limited. One option, often reserved for the youngest son, was to join a monastery. In some rare cases, men would move to Karja (*Gar zha*) to take up farm or labor employment.[24] Another option was to become a *makpa* (*mag pa*), a status acquired through uxorilocal marriage, in which a man takes up postmarital residence in the household of his wife.[25] The remaining option, and that most commonly adopted, was for men to join their eldest brother in polyandrous marriage. Polyandry was once highly prevalent in Ladakh: in the words of many elders, in the past "everybody was living like that."[26] Yet according to Ladakhi elders, it was sheer poverty that compelled people to live in polyandrous marriage; for this reason, the 1941 ban on polyandry is rarely cited as a source of changing practice, but rather improved access to employment opportunities with the economic restructuring of Ladakh after the Indian independence.[27]

Ladakhis often describe polyandry as a system of cooperation, and polyandrous marriages indeed are akin to small cooperative enterprises.[28] Polyandrous unions are often described by elders, by both men and women, as *skitpo* (*skyid po*), or pleasant, precisely because work was shared (see Bhan 2014:35; Hay 1997). But polyandrous unions could also lead to tensions among brothers, especially because some elder brothers were authoritarian: the eldest son of a family was, in the words of many, "like a king," and the lot of his brothers depended on his degree of authority.[29] In cases where such tensions could not be resolved, the remaining options were again very few, especially because in Ladakh having various married family units under the same roof was widely considered unacceptable (Crook and Crook 1994:742). It is in this context that in postindependence India employment in the army and in the civil service opened new opportunities outside the agrarian system and an escape from this household organization, which could be experienced as a rigid structure.

"Elders in Ladakh say that lots of construction going on is a bad omen, that bad times will come," continues the *onpo* in a grave tone. "Look around. There are so many new roads being built. Are all of them really needed? All this machinery and these cars are just destroying everything," he laments. "And there are so many new houses, fancy ones, some of which are left empty for so many months of the year." With amusement he notes, "I have heard this is the case in Leh, that people are being paid to take care of other people's homes while they are away, just to make sure they don't get fined for not fulfilling their share of community work."[30] The *onpo* is quite right. A military man living not too far from Namgyal's home in Leh who is posted far away, where his wife had followed him, is indeed paying a woman to stay there and take care of the family's share of community work. Such an arrangement illustrates significant changes in a society where community work has long been a key aspect of the organization of life. It equally reveals not only the profound moral obligation toward community responsibilities, but also the potential of social stigma associated with the failure to fulfill them.

"Ladakhis have for generations faced hardship," the *onpo* observes. "Look at this dry land. See how cold it is in winter. And the stories our *abi meme* would tell us all about the abuse of these people from Khachul," referring to the exploitative life conditions under Dogra rule. "Ladakhis were starving! The only way we could survive was by standing together. We had nothing, but we had each other." As if to confirm his point, the man's granddaughter, a toddler with plump rosy cheeks, arrives in tears, having stumbled on a rock. She climbs onto her grandfather's lap in search of comfort. The man consoles her and continues, "Today, the more people are building, the more they forget about each other, even their own father and mother." This is a leitmotif in descriptions by elderly Ladakhis of life under the Dogra/British administration before the independence of India: amid hardship and abuse, and with the exposition to the everyday vagaries, community cohesion was a moral imperative. There were obviously abuses within Ladakhi society itself, as Tsewang Jorgais points out with the case of the moneylenders, and even within families, especially when elder brothers adopted an autocratic attitude. However, elders uniformly agree that maintaining cohesive community and family relationships was fundamental.

But there is more than a changing attitude toward one another in the villages today, as the *onpo* suggests. There are objective circumstances that contribute to fragmenting social relations and elders' vulnerability. The deep collective fracture evoked by the *onpo* has concrete roots that can be found in the fundamental economic unit that constitutes the Ladakhi household. The extended family has traditionally provided a social safety net for aging people; however, the reorganization of the Ladakhi household structure now makes this safety net ever more tenuous.[31] Over the past three decades, rural desertion due to outside employment has resulted in an increasing number of elders living alone. Only a few decades ago, two elders having to battle everyday matters and the toll of aging on their own, like Morup and Diskit, would have been unusual. As Childs (2004:129–41) eloquently demonstrates, and elders of Ladakh are keenly conscious of, aging in the Himalayas today often means becoming someone else's burden. I often heard elders who have become dependent on co-resident sons and daughters-in-law lament their inability to help in the household and the farm due to their old age, and their consequent feelings of guilt. This sentiment is further aggravated when household context does not allow elders to contribute to its productivity through other forms of valuable work, such as the passing on of knowledge to younger generations or the caring for young children.

However, as much as they would like to support their aging parents and in-laws to the best of their abilities, as Angmo was doing for Morup and Diskit, many Ladakhis are constrained by the imperative of ensuring the subsistence of their young families, which takes them far from their natal village for extended periods of time. As Childs (2004:132) remarks, in the Himalayas "the cultural ideal of reverence for the elderly is mitigated by the harsh pragmatism of economic reality." With access to paid work, Ladakhis have seen a steady trend toward life in nuclear household units and an associated increase in the partitioning of family land estates. Land lots in Ladakh have always been relatively small, but their further fractioning and lack of support for pastoral work make it increasingly difficult to sustain livelihoods strictly from farming.[32] The new personal freedom brought by paid work also leads to new expectations of the good life, which is no longer primarily defined by communal values. Amid changing values, the agrarian and pastoralist lifestyle is now leaving many aspirations unfulfilled.

CONCLUSION: FROM REMOTENESS TO LONELINESS

Life under Dogra rule in preindependence Ladakh was characterized by the extreme psychological and physical abuse of Ladakhi peasantry. While accounts of life in those days often revolve around tropes of hardship, whether in the form of extortion or starvation, in an apparent contradiction Ladakhi elders concurrently remember this past as a happy one, marked by a meaningful collective life, in contrast to the present-day shattered sense of purpose. If recollections of the past may at times idealize the cohesive nature of past communal and family life, such perceptions are largely nourished by a present sense of anguish, quite tangible among many Ladakhi elders, created by the shift from being a productive to a dependent member of the household. Thus, although the autocratic rule of the Dogra has been abolished and the government has brought development, the present, characterized by a feeling of loneliness and a growing sense of anonymity, seem inauspicious.

In the current context of Ladakh, however, it is increasingly difficult for people to ensure their well-being or fulfill their aspirations through traditional subsistence activities. Moreover, because employment and education opportunities are generally available only outside the villages, family members are forced to leave their villages for extended periods. Doing so is often the only way a family can ensure their subsistence and enable its members to fulfill their aspirations. If, as elders suggest, younger Ladakhis neglect the moral duty of caring for one another, particularly for family members—a duty that has enabled generations to survive both the vagaries of life in a harsh and unpredictable environment and the abusive governance of the Dogra—the fact remains that the younger generation is facing drastically different life circumstances, which bring their own dilemmas.

If there is something universal in different generations, whether looking forward or backward, feeling that another generation has or had it easier, Ladahkis' experience of life has also objectively changed between generations. Today as in the past, old age is fraught with difficulties. For Ladakhi elders, this inescapable fact of life is exacerbated by an eroding social security net as the household becomes less and less a site of generational integration. With the steady shift from joint to nuclear family arrangements, ancestral land is progressively divided and family lands splinter into anonymous fragments.

Many see in these developments the sad ending of a life organized around communal and family integration. As Tsewang Jorgais elegantly put it, "Although we had lots of land, we had just a single apple tree. Later, as land got divided, even these apples went away from us."

The growing difficulties associated with sustaining agriculture and pastoralism constitutes another dimension of the changing nature of the household as a site of integration. Agro-pastoralist activities are profoundly social and hinges on community and family arrangements. Yet these arrangements are today eroding under demographic and economic changes. The decentering of the traditional economic activities of Ladakh cannot be isolated from the politics of nature that has prevailed in Ladakh since Independence and by which the region has been reconfigured as a fortress for India. This politics is a product of the postcolonial geopolitical context of India, which has changed the faith of Ladakh by throwing the region into the heart of one of the most sensitive and still-unresolved conflicts of contemporary South Asia. It is a politics with roots in the Partition of India, which continues to shape how a generation of Ladakhis reflects on environmental changes today.

CHAPTER TWO

Arthalis and Beyond

A Crack in the Landscape

In her country, do they have the fear of Pakistan?

—DOLMA YANGZOM, NYEMO VILLAGE, IN REFERENCE TO ME

By the grace of lord Buddha we won the war against the Balti.

—LHAMO PUTIT, SKINDEYANG VILLAGE

Before sleeping
There are some questions to be answered
Before dying
There are some statements to be made

—TIBETAN PROVERB

IT IS SUMMER 1948. TWO YOUNG WOMEN FROM NYE ARE GRAZING their cattle on an alpine grassland when they see a group of tall armed men dressed in black approaching from the upper part of the valley. Curious about who these men are, the women have no idea they are affiliated with a group that is playing a crucial role in the first Indo-Pakistani war. A few months earlier, in 1947 when India became an independent nation, Jammu and Kashmir was a princely state, formed by a conglomeration of independent kingdoms, among them Ladakh. After an initial hesitation, the Maharaja of Jammu and Kashmir pledged allegiance to India. A few months after, in February 1948, Pathans[1] and Gilgit Scouts, a military group that the British had organized to defend themselves against Russia during the Great

Game, and that pledged allegiance to Pakistan, attacked Skardu in a surprise invasion of Jammu and Kashmir that eventually led to war between India and Pakistan (Gutschow 2006:473). The chaos of the partition had reached even the high terrain of the Himalayas. The hostilities that began in 1947 affected Ladakh in 1948.

These political developments were largely unknown to Ladakhi villagers until the news spread throughout Sham that Pathans and Gilgit Scouts were roaming the highlands pillaging and looting everything in their path. The men were marching toward Leh, intending to take control of the Ladakhi capital.

It is one of these groups that the women herders meet in Nye, less than 40 kilometers from Leh. "Where do you come from?" asks one woman in typical Ladakhi salutation when the two groups meet. "Are you the Indian army or you are Balti"?[2] Unsuspecting, the young women bring the raiders to their village. (In those days, encounters with both Indian infantrymen and colonial administrators were rare for villagers.) When they reach the settlement, a villager stops them. "We were expecting you and so we had made tea for you," he says, also thinking that these are members of the Indian army. Aware of how tiring mountain passes are, Ladakhis are very hospitable.

From a hill beneath the village, Tsetan Zangpo witnesses this interaction and shudders with fear. He knows who these men are. Like many Buddhist Ladakhis, Tsetan Zangpo has recently been assigned by the Indian army to guard his village. Amid the turmoil, Indian forces rely on Buddhist Ladakhi civilians for their expertise in the territory, organizing them into a local ethnic guerrilla force. From his vantage point, Tsetan Zangpo can also see toward Nyemo, one of the last villages before Leh, where a large Indian army contingent is stationed. Taking advantage of time afforded by the villagers' gregarious greeting, he hurries to the upper part of the village. Feigning ignorance, he tells the visitors that the Indian army has just left and directs the Pathans towards a trading road leading to Nyemo. Once the Pathans have finished their tea, their contingent heads in the direction Tsetan Zangpo had indicated. This is the opportunity he is hoping for: as the raiders make their way down the road, a group of villagers takes a shortcut, using the mountain paths they know so well, to warn the Indian troops at Nyemo of the approaching danger.

Tsetan Zangpo now recalls with the calmness of his ninety-two years, "The Pathans went straight toward the Indian army." The opposing forces clash over the next few days. Because Tsetan Zangpo and his fellow villagers warned the Indian troops, the latter were able to ambush and defeat the Pathans. Tsetan Zangpo's recollections provide the backdrop to Indian state formation in Ladakh. Until the independence of India, the physical presence

CHAPTER 2

of the colonial administration remained marginal in the region today known as Ladakh, which was governed from Jammu and Kashmir by absentee Dogra rulers. The region was also militarily vulnerable: at a crucial stage in the conflict, when Kargil fell to the raiders in early May 1948, only thirty-three men from the State Forces[3] remained in Leh to defend the vast territory of Ladakh (Verma 1998:24–25). When the raiders invaded their land, Ladakhis were at their mercy.

That Leh did not fall into the hands of the raiders is something that confounds military analysts. In their study of the military operations in Jammu and Kashmir in 1947–48, commissioned by the India's Defense Ministry, Prasad and Pal ([1978] 2005: 346) reflect: "There appears no doubt that in the last week of May 1948, and again in July, Leh was within the raiders' grasp. It still remains a mystery why they did not press home their attacks and capture Leh once for all. . . . It is true that the local villagers were against them, but their opposition was meek and entirely passive, and it was of negligible importance in that vast, sparsely populated and thinly held region."

This perspective blatantly ignores the active role of the local population in determining the outcome of the conflict in Sham. The Indian army's defensive task in Ladakh was daunting, and its success depended on the mobilization of the local population. Many Shammas provided support for the Indian troops by taking up arms. Testimonies from men and women who experienced the war suggest that the support of the local population was decisive, both for the outcome of the war in Sham and, by extension, for the history of postindependence India. When Ladakhis speak of the first Indo-Pakistani war today, they speak of *arthalis*—the number forty-eight in Hindi—to distinguish it from three later wars India fought with Pakistan (1965, 1971, 1999), all of which partly played out on Ladakhi territory as well.

The first Indo-Pakistani war was a "critical event" that shattered local worlds (Das 1995). Not of the Ladakhis' choosing, the war was a source of ethical dilemmas that for many remain unresolved. The politics of religious identity as it played out during the war, coupled with Shammas' fear of losing their land, led them to ally with the Indian state. In retrospect, although joining the Indian army may have appeared to be the only option Buddhist Ladakhis had, many informants spoke of their initial reluctance to take active part in the conflict. There were several alternatives open to Shammas at the time—ethical neutrality, alliance with India, or even befriending the invaders—and in the simplifying fog of war it was the politics of religious identity and a sense of moral responsibility that inspired them to defend their land alongside the Indian army. For as many informants confided, over several months the prospect of India's victory seemed bleak.

Ladakhis live uneasily with the tormenting memories of the war, which many see as invested with potential karmic retribution. The distinction between memory and history has been the object of much scholarly scrutiny. In an effort to go beyond the notion of "false memory," Antze and Lambek (1996) have studied memory as an interpretive reconstruction structured by local narrative conventions and cultural beliefs. For historian Pierre Nora, history is the ever-problematic reconstruction of the past, of "what is no longer," whereas memory is a living thing: it is embodied in individuals alive today and, therefore, "remains in permanent evolution, open to the dialectic of remembering and forgetting . . . a perpetually actual phenomenon, a bond tying us to the eternal present" (Nora 1989:8). My point in emphasizing the subjectivity of the war is to understand why, as a "critical event," it may have implications for the karma of Ladakhis, as many elders suggest. What is of interest here is therefore not so much the precise reconstruction of a historical event, but how the war shattered the worldview of those who lived through it and how those individuals reflect on associated events. Shammas' reconstructions of the war are culturally mediated, and they emphasize the moral wrongs of the event. Taking this into consideration helps shed light on their interpretations of a changing Ladakh today, which many elderly Shammas see as having something to do with the war.

"MAY SUCH A THING NEVER HAPPEN AGAIN"

Though largely glossed over in the history of postcolonial India, the fact is that during the first Indo-Pakistani war Ladakh nearly fell to raiders from Pakistan. For those in the occupied villages who eventually found themselves in clashes between the Indian army and the invaders, the year 1948 is suffused with painful memories.[4] Commenting on how the invasion had completely disrupted village activities because people were living under the constant shadow of fear, an elderly woman said to me, shaking her head, "May such a thing never happen again." In Sham, at the climax of the threat to Leh, which lasted from July to October 1948, raiders were positioned in the villages of Sumdha, Saspol, Basgo, Nyemo, Umla, and Taru.[5] Elders who lived in the occupied villages refer to this period as "when we were under Pakistan." The stalemate continued well into the fall, neither side gaining a decisive advantage until early November 1948, when Indian forces made strong advances, reaching as far as Marol in Baltistan later that month. By late December, heavy snow brought all operations to a standstill. While the opposing forces remained in their respective positions, political negotiations finally resulted in a UN-mandated cease-fire that came into effect on

January 1, 1949 (Prasad and Pal [1978] 2005:343–45). Ultimately, the defense of Leh lasted almost a year, from February to December 1948, a period during which the population of Sham lived in fear, under constant threat to their land, their home, and their identity.

The raiders supplied themselves with booty stolen from the villagers. Many captive Buddhist Ladakhi men were forced to work and regained their freedom only when the conflict ended. One of these captives was Ishey Wangail from Nyemo village. His story represents what many Shammas went through.

When Ishey Wangail hears that the invading forces have penetrated Sham in their march toward Leh, his first concern is to save his family's cattle. News has traveled as fast as the wind that summer, and he has heard stories of the raiders indiscriminately slaughtering animals to feed the troops in every village they passed through and forcing villagers to work as porters. In May 1948, when the conflict reaches a critical stage, Ishey Wangail decides to guide his cattle into the mountains, where the animals will be safer, while he continues on to the village of Umla to take refuge in a relative's house.

After a few days, wracked with worries about his family and his cattle, Ishey Wangail makes his way back to Nyemo with two fellow villagers. On the way, however, they are captured by a detachment of hostile forces. Seeking animals for transport, their captors first bring them to Nyemo, but when they reach the village they find it almost empty: panicked, most villagers had fled to the mountains, as is the case with villages throughout Sham. But although many of the normal patterns of life have been brought to a sudden halt, farmers in Sham have no choice but to carry on with their farming activities: they depend on their harvest to survive the winter. In many households, only the elders and one able body have stayed behind to look after the fields; most often it is the women who stayed behind, as the men are either fighting alongside the Indian troops, being held captive for forced labor, or hiding in the mountains. This forces women to live with the raiders, many of whom have taken up residence in village households.

After their captors seize what little cattle remains in Nyemo, Ishey Wangail and the other detainees are taken toward Khalatse. Over the next several days, the group is forced to carry loads of vessels, dung, wood, food, and jugs of water for the raiders in the mountains. Like many other Ladakhis, Ishey Wangail refers to this work as *begar*, a term that unmistakably connotes the forced labor of the exploitative, pre-Independence rural economic system.[6] Escape is near-impossible, as detainees are under constant surveillance and, when not moving, are tied to heavy stones. But Ladakhi prisoners offer a degree of resistance: narratives abound of detainees not only escaping, as Ishey Wangail finally did, but also throwing arms, ammunition, and other equipment into streams.

The villagers must provide for the raiders in various ways. In some places, every household has to provide twelve goats as rations. Many women are told to wear their *perak* (*be rag*)[7] and to perform dances. But the women are not wholly compliant. For instance, in Nyemo the raiders occupied the house of Skalzang Yangzes's in-laws. With them was a Muslim Ladakhi from the area of Purig who had joined the raiders. He boasted to Skalzang Yangzes's mother-in-law that they would soon reach Leh and buy gifts for her, as no Indian army would stop them in their conquest. Skalzang Yangzes remembers her mother-in-law bravely confronting her persecutor, telling him: "Whoever may die [because of this situation], let it be so!"

At one point, Ishey Wangail and two fellow villagers are forced to carry loads toward Khalatse. Food is scarce and they are starving. Ishey Wangail worries about his hungry horse, too. The raiders tell them they will be set free in Saspol but that their horses will be kept until they reach Khalatse. Because the prospect of losing a horse is bleaker than remaining captive, all three men decide to continue toward Khalatse. However, Ishey Wangail soon realizes that the raiders will likely never let him go. And in this, he may well be right: many Ladakhi captives regained their freedom only when the UN cease-fire ended the conflict, and had to make their way home in the middle of winter from distant Baltistan.

When they near Khalaste, the three men escape into the mountains at the first opportunity, eventually reaching the village of Hemis Shukpachan. There, they learn that more raiders are approaching from multiple directions, further complicating the Indian army's defense. The group of escapees briefly stops their flight to perform *storma*, an exorcist offering to ward off evil influences and danger.[8] They then continue their trek toward Likir under the cover of night, and since they encounter no sign of the raiders, they decide to press on toward Nyemo. Reaching Nye in the darkness, they are relieved to encounter villagers gathering their animals for use by the Indian army: reinforcements have finally arrived. At long last, there is hope of driving back the invaders.

ORGANIZING THE RESISTANCE

By early May 1948, many Ladakhis see the prospect of losing their land as increasingly possible. Busy fighting in the Kashmir Valley, high-ranking Indian army officers have been reluctant to spare any men to help defend

Ladakh (Chibber 1998:144–50). Word of the invasion of Ladakh comes to Srinagar and reaches Prithi Chand, an officer from Lahaul with family and religious ties in Ladakh. The attachment and feelings he has for the region, as recollected in his biography, compel him to volunteer to lead the defense of Leh. Prithi Chand and his small detachment of fifteen men, most of them Lahaulis, trek through harsh and perilous conditions for three weeks in brutal cold and snow before reaching the Ladakhi capital. To Prithi Chand, their passage through such obstacles feels like a miracle (Chibber 1998:150). In Leh, they proceed to raise militias from among the local Buddhist Ladakhis, who over the next two weeks undergo basic training. The hastily assembled defense force has little equipment and many Ladakhis are initially armed with only traditional lances and bows (Prasad and Pal [1978] 2005:327–28). This is how sparse the defense of Ladakh is in the first weeks, before reinforcement finally arrives at the end of May. As Malhotra (2003:6) points out, these Lahauli officers would play invaluable, if largely unheralded, roles in defending the new Indian republic's territory.

The decision to join the militia is, however, not instinctive, and many Buddhist Ladakhis are initially uncomfortable participating directly in acts of violence, something that can only bring retributions. Although none of my interlocutors reported cases of Ladakhis fighting against the Indian army, many confided that they left the army as soon as they could, to assuage the mental strain caused by this work. Torn between the need to protect their land, moral prohibitions, and, as many made clear, the fear of dying, the answer to the call to take arms was agonizing for many recruits. In the beginning, in fact, a system of conscription was put in place; in some villages, each household was asked to provide one able-bodied man to support the Indian troops. But this situation did not please everybody. As a man from Basgo who was conscripted puts it, when young men were given arms and exhorted to go fight in Khalatse, "some went crying and some went singing." Clearly, not all Ladakhis wholeheartedly embraced work with the army, and many would have preferred not to align with any group. There were also incidents of coercion and physical punishment. Shedding tears, one man from Hemis Shukpachan reports having been tied up and beaten by Prithi Chand himself for having his gun stolen by the raiders.[9]

Many residents of Sham provided support to the Indian army without joining the ranks. Ladakhis' knowledge of the territory, acute sense of the mountains, and familiarity with the paths crisscrossing the region proved invaluable in defeating the Pakistani invaders. Many of those who fled to the mountains to escape the raiders volunteered to guide Indian troops through the mountain trails. Others lured invading forces into Indian army ambushes, acted as spies and lookouts, or lit fires at night to distract the enemy. Many Ladakhis were asked to obtain information about the raiders, and some would act as runners to communicate information. Villagers attempted to deceive the invaders by making fake tracks along mountain trails, creating the impression of large Indian army contingents moving through the region. They also provided fundamental logistical support, notably as mountain porters, and village headmen provided the Indian army with food, woolen clothes, and other goods (Chibber 1998:159–60).

The testimonies I collected make it clear that many Ladakhis were ambivalent about whether or how they should take part in the conflict. Some were pressured by their family members and fellow villagers. Others saw a pressing need to actively resist the raiders. It is nonetheless clear that a great majority of Shammas supported the Indian army, and this may have to do in great part with how the politics of religious identity played out during the conflict.

FRAUGHT RELATIONSHIPS AND
THE POLITICS OF RELIGIOUS IDENTITY

The politics of religious identity took various forms during the conflict, including crude insolence, and led many Buddhist Ladakhis to believe that Buddhist-Muslim coexistence would have been problematic had Pakistan won the war. In Hemis Shukpachan, for instance, raiders reportedly denigrated Buddhists for their customs, calling them illegitimate children (*nyal bu*) in reference to the practice of polyandry. Buddhists were also prevented from performing religious rituals. In Saspol, villagers were gathered together, forced to listen to raiders' speeches, and challenged to raise their hand if they believed that Buddhism was great. In Skindeyang, the raiders made all the men and women gather, tied the men's hands, and forced them to shout, "Long live Pakistan, death to Hindustan!" For many, this unquestionably indicated that the invaders were determined to transform Ladakh into a "Muslim land."

There were also acts that upset cultural ideals. The shock created by the slaughter of animals during the occupation cannot be overstated. In often quite emotionally charged testimonies, most of my interlocutors expressed their sadness at seeing animals being killed in massive numbers.[10] Ishey Wangail (whom I introduced above) recalls the traumatic memory of seeing on a mountainside an enormous stack of bones and other remains of yaks and *dzo*, all that remained of the cattle the raiders had killed and eaten. "They were stacked up like horns on a *lha tho, konjok!*[11] It was unbelievable," he murmurs, still disturbed by the scene. But what truly broke his heart was to discover, on his return home to Nyemo after weeks of captivity, his entire herd of yaks, *dzo*, and cows butchered, their bones littering the ground. "They were all slaughtered by these merciless people," Ishey Wangail tells me in tears, his voice trembling. "I have gone through tough times and seen hard life that no one should have to see." The slaughter of animals during the occupation was clearly a painful memory for many Shammas, the raiders' purely utilitarian approach to livestock jarring their consciousness.[12]

If the reasons behind the invasion of Ladakh remain unclear,[13] many people in Sham believe that the goal of the raiders was access to land, an assumption that calls on long-standing perceptions of cultural practices. Buddhist Ladakhis see polyandry as a custom that preserves land assets. It is often justified by comparisons with the agrarian situation of Muslim Ladakhis, who have fewer assets, something Buddhist Ladakhis attribute to the splitting of land among brothers. In fact, during the war, the Shammas were surprised to recognize familiar faces among the raiders: people from the "Balti rong," an area on the way to Kargil, also known as Purig, had joined the raiders in their march toward Leh, guiding them and acting as interpreters.[14] Shammas remember how they arrived in the villages pretending to be saviors, saying *"assalamu alaykum,"* the cordial Arabic salutation adopted by Muslim Ladakhis, shaking hands and inquiring whether the villagers had been troubled by "Hindustan."[15]

What drove some people from Purig to sympathize with the raiders may have roots in existing equivocal relationships between them and the Shammas. Interactions between Buddhists and Muslims have always been an integral element of the social landscape of Ladakh, as evidenced in trading, among other activities. Former traders' testimonies suggest that the sociality that characterized trading relationships differed from Tibet to Muslim areas (see Rizvi 1999:95). Trade in Gertse (Tibet) was characterized by relationships of mutual affection, loyalty, and honesty.[16] Despite the various potential dangers along the way, including banditry, former traders all described their trading activities in Gertse in passionately positive terms.

Their experience when trading toward Baltistan, however, was far less affectively rewarding. Many alluded to the challenges of trading in Kargil, including being called names like *ro bodh* ("dead Buddhist bodies"),[17] having mud thrown at them by children, being teased by youths riding traders' donkeys, and being prevented from entering people's property because they were considered dirty and polluting. Accounts of trade within Sham itself also suggest fragile relationships. Muslim Ladakhis from Chiktan village, in the area of Purig, would regularly trade to Sham, exchanging vegetable seeds, turnips, and other minor barter items for wooly wastes with which they would weave clothes. Villagers gave them shelter and food, and they would also take odd jobs, clearing farmland and building houses. People from Chiktan were renowned as storytellers, and Buddhist hosts would stay awake late into the night listening to their Muslim guests' stories, an image that evokes rather congenial relations.

Although religion seems not to have been a source of tension between the travelers and their hosts, there was an awareness of inequalities. The people of Chiktan traded and worked in Sham because their homeland was comparatively poor. This is also what brought them to Sham during Losar, in order to impersonate characters in the various dramas and performances that took place during the festivities. Undoubtedly, they would take these odd jobs out of sheer poverty, as these enactments were sometimes demeaning: for instance, in some villages, the personification of the character known as the *lama dzugu* required jumping into a stream in the middle of winter. The amused tone with which some Shammas recollect their past interactions presumes a lack of empathy, by some, for their visitors' desperate and destitute circumstances. Shammas also emphasize the indigence of the "Baltis" from Chiktan, a situation they invariably attribute to a lack of land.[18]

Amid this social context, for the Muslim Ladakhis from Purig area, the war had the potential to reverse old inequalities and raise their status in relation to Buddhist Ladakhis.[19] Many interlocutors insist that people from the Balti rong were crueler during the war than the Pathans; many (though not all) even depicted the Pathans as "good at heart" and less abusive. One of my informants from Khalatse remembers how the Pathans would say that they were just military men who had to fight and that this fight was between them and the Indian army, not the population. However, it was also not uncommon for men from the Balti rong to simply inform villagers that their land and houses now belonged to them.

Overall, these dynamics suggest that in Sham the war took on a specific communal tone leading to fraught relationships between the people of Sham and Purig.[20] For the Shammas, the occupation was a period of turmoil, the potential loss of the war a threat to cultural and religious identity.[21] The fear of this scenario is tangible in the testimonies of those who lived through the war, and I frequently encountered the idea that Ladakh would be a "Muslim place" today had it not been for the intervention of the Indian army. Reflecting on the motivations for enlisting, one man from Nye tells me: "It was done for our land, for our families, and to protect everything." Many believe that, had the conflict turned in favor of the raiders, belonging in Ladakh would no longer be the same. "We could have lost all our monasteries," says one

CHAPTER 2

man, alluding to the rationale for enlisting in the militia. These reflections echo the thoughts and experiences of many Shammas.[22]

The Shammas sense of moral responsibility toward defending the land and religious institutions may also have been nurtured in part by the discourse of the Lahaulis. For instance, Manohar Lal Chibber (1998:153–54) reports the following speech given by Prithi Chand in Leh on March 13, 1948, while trying to recruit Ladakhis in the militia:

> As you know, all the passes are closed and it is difficult to get
> reinforcement to this place. Till such time the Army arrives,
> we have to defend Ladakh at all costs. . . . All of you know that
> in Pakistan thousands of innocent Hindus and Sikhs were killed
> and their temples burnt. . . . Now if you wish to protect the mon-
> asteries, save our culture and the honour of our women and reli-
> gion, then come forward, get training in arms from us and join
> us to fight the invaders. . . . While I request you to maintain har-
> mony, I warn you that whoever sympathizes, propagates or helps
> Pakistanis will not be spared.

This rhetoric, which, given the ongoing turmoil, may have led some to interpret the war in communal terms, reached far beyond Leh. In Sham, the language of a common genealogy underpinned the mobilization of Buddhist Ladakhis. The Lahauli officers arriving from Srinagar appealed to Buddhist unity and Himalayan imagery in recruiting Ladakhi volunteers. One of my interlocutors remembered how in Nye, Prithi Chand gave a speech to the villagers in which he said, "People of Ladakh and people of Karja [Lahaul] are under the same star, and therefore it would not be right for us to leave you all to die here. That is why I came here and brought some men along with me." Idioms of kinship, blood, and family are potent rhetorical strategies for instilling loyalty to the nation (Borneman 1992; Herzfeld 2005:2). Such ideas must have traveled fast in Sham that summer. This genealogy was apparently decisive in attracting villagers' sympathy and confidence; in fact, in our conversations, the Shammas who lived the war recurrently evoked the bonds between the Lahauli military group and Buddhist Ladakhis.[23] This rhetoric may have gathered strength as the war played out along communal lines in the occupied villages and, in its emphasis on a Buddhist genealogy, only reinforced communalism, thus ignoring prevailing social and kinship bonds between Buddhists and Muslims in Ladakh.

Ladakh has a rich heritage of cultural hybridity and intercommunity marriage, but as scholars have shown, a chasm has grown between Buddhists

2.1   Makeshift tomb of a Pakistani soldier who died near Yangthang
       during the first Indo-Pakistani war

and Muslims over the past decades.[24] My interlocutors' memories of the war
may well be mediated by current tensions. Certainly, the situation may not
have been without strain before the war either, as the accounts of trade above
suggest. But rather than religion per se, in those days ethnicity, place of
origin, and social status were more likely to differentiate communities

(Demenge, Gupta, and Deboos 2013:9). In fact, if the context of the war nurtured communal tensions, this communalism did not appeal to everybody.[25] On several occasions, Buddhist interlocutors shared their disapproval of attacks made by their fellow villagers on some "Baltis," whether revenue officials, traders, or travelers, who happened to be in Sham when the war erupted. Clearly, some Buddhist Ladakhis were reading the conflict as a communal one, though this perception was not shared by all. Today, a generation of Shammas continues to live uneasily with this past event. Memories of the war, whether embodied or inscribed in the ruins that still punctuate the landscape, shape how Shammas reflect on current changes in the environment.

WEATHERING A STORMY PAST

In Basgo village, up on a craggy clay mountain, stands the Rabstan Lhartse fort. Today a crumbling ruin, in the 1680s the palace withstood a three-year siege by Tibetan-Mongol forces. When King Delek Namgyal could no longer hold out, he appealed to the Mughals for assistance, who sent reinforcements from Kashmir and Baltistan, and they defeated the invaders (Halkias 2013). Nearly three centuries later, Basgo once again witnessed a Ladakhi stand against foreign invasion. On the side of the road that runs below the fort stands Hemis Labrang,[26] a building where a decisive event in the denouement of the first war between Pakistan and India in 1948. That on this occasion the forces of Baltistan were the invaders is testimony to the shifting alliances that have marked Ladakh's history.

One day in February, Namgyal and I are sitting next to this site, as yet unaware of its role sixty-five years earlier in the partition of India, even though it is the question of the Partition that brought us to Basgo in the first place. The name of the village has come up repeatedly in elder's accounts, for it was here that Hari Chand, a Lahauli officer, destroyed a 3.7-inch howitzer, a mountain gun brought by the raiders. Many Shammas feel that the cannon's destruction allowed the Indian troops to repel the raiders, marking this event as a turning point in the conflict. However, for reasons I cannot work out, people in Basgo are not as keen to discuss the Partition as others are elsewhere in Sham; they seem almost uneasy. This puzzles me, since thus far Ladakhis have been amazingly generous in answering my numerous queries. What is more, discussing the events of the Partition seems to have a welcome cathartic effect for many. But in Basgo, things feel different. I am beginning to feel like an intruder in this community. "Don't you feel there is something strange here?" I ask Namgyal. "There's a strange vibe, as if people aren't comfortable talking about the war, don't you think?" Namgyal is just as perplexed as I am. "I can clearly feel something is strange," he confirms.

Our musings are interrupted by an elderly woman who approaches us slowly, a cane in one hand, a prayer wheel in the other. "Where are you coming from?" she asks. We introduce ourselves and explain the reason for our presence, though I am careful to avoid mentioning the war. My reference to environmental changes in describing my project inspires her, however, and she begins talking about the water situation in Basgo and how badly the village was affected by the flood that took place in 2010.[27] Then, in the same breath, she tells Namgyal: "*Nono*, you must be a Muslim from Leh, as you are soft-spoken." Namgyal is always careful to speak in a calm and composed manner when we work. In the villages, people quickly peg him as an urbanite from Leh, given his style of dress, refined manner of speaking, and obvious lack of practical knowledge of certain aspects of Ladakhi village life. But this is the first time anyone has asked about his religious identity, and, more intriguingly still, the woman clearly lowered her voice when saying the word "Khache" (Sunni Muslim). "I am from Leh and I am Buddhist, *abi-le*," replies Namgyal, visibly stunned by the question. "Is that true!" she exclaims. "The younger generations today no longer wear the traditional dress," she complains, implying that that would have made Namgyal's identity much easier to ascertain.

Sensing an opportunity, Namgyal asks me, "Shall we ask her about the war?" Without giving me a chance to reply, he proceeds to ask the woman if she was in the village during the war. "Of course I was!" she exclaims. Lowering her voice, she points to the Hemis Labrang building next to us: "This is the exact spot where the dangerous cannon was," she explains, hidden and guarded by Pakistanis. "These Pakistanis, they had taken some villagers and had forced them to act as watchmen outside the building. When Hari Chand and his men attacked, those local men did not warn the Pakistanis inside," she says hastily. In fact, they did the opposite: They indicated the emplacement of the cannon to the commandos. Except for one man who managed to escape, all the raiders inside the *labrang* were killed when Hari Chand's commandos attacked. "These poor men," says the woman, a grave expression on her face. She then points to the gate of the now vacant building: "And then, as if it was not enough killing, the villagers [who were hostages] later got executed by the raiders, on the bank of the Indus," she says, visibly holding back tears. The events clearly still weigh on her after all these years.

The assassination of fellow villagers was the source of much consternation in Basgo, where people had been at the mercy of the raiders for weeks. "I still remember the gunshots," another elderly woman confides later on, her fingers busy with the beads of her rosary. "I can't remember everything, but it is not even worth sharing," she adds, implying that these events were deeply unsettling. Ishey Wangail, whom I introduced earlier, saw the execution of the villagers by the Indus unfold in front of him during his captivity and is manifestly still disturbed by the incident, confiding in a trembling voice, "One of them tried to escape by jumping into the river, but he was shot in the water. I can still see how the blood colored the water red."

As she shares these painful memories, the old lady stresses that she did not see all this happen with her own eyes, for she was grazing sheep and goats in the pasture between Basgo and Likir at the time. The *goba* ('*go pa*), or village head, had dispatched a few villagers to stay in the pastureland with animals to cover the footprints of the commandos and facilitate the men's escape. She did, however, see Hari Chand and his troops as they headed toward the Nubra Valley, making an impressive detour toward Leh to avoid the raiders. "The Indians went off at noon," she says, adding: "the Buddha must have been guarding them, as they were going in broad daylight." That Hari Chand seemed to have had the blessing of divine power is a widespread perception among the Shammas who experienced the war.

But the woman's thoughts quickly return to Namgyal's identity. She stares at him for a moment, then remarks: "You don't have any sign of a Buddhist." She laughs and continues: "I am being very true to you, I am saying all these things, and if you are a Buddhist, you should show me a sign that you are." "I am Buddhist, I am Buddhist, *abi-le*," repeats Namgyal, stupefied. She laughs, then observes, "Right now, I don't know what to make of you. You don't even have a *tawiz*," referring to the religious locket worn by Muslims. The growing anonymity in a place where everybody used to be easily identifiable is something that feels at times distressing for a generation of Ladakhis like her. "Do not lie to me, I am being very frank!" she exclaims. Namgyal stays mute, now manifestly vexed. At this, the woman leaves, explaining that she has other matters to attend to.

• • •

The violence of the war left deep wounds and sowed seeds of long-standing communal tensions. When the raiders retreated before the advance of Indian troops, amid chaos and fears of reprisals, the Muslim families of Sham abandoned their homes and followed them to Baltistan.[28] After the events of the Partition, travelers from Chiktan stopped coming to Sham for barter and work. According to my interlocutors, relations between Buddhist and Muslim Ladakhis were clearly different after the war and were soon mired by conflicts.[29] This explains the elderly woman's feelings of suspicion regarding Namgyal's identity, something he is still having a hard time coming to terms with. Later that day, as we are sitting at the edge of a cliff overlooking the Indus, which bounds the village, Namgyal asks, "Why was she picking on me like that?" "She was not really picking on you," I suggest, unable to find the words to make him feel better. "It must be the context, Namgyal. She was discussing very sensitive issues." This was not the first time Namgyal had come across as an outsider. During our work together, he was often teased for his lack of knowledge about farm-related matters or the local vernacular—the toll of the years spent outside the region for his education. Generally, he was not troubled by it. But this time, the persistence of the woman's questioning clearly unsettled him. "Why didn't you tell her your house name, which family you are from?" I ask him, for this time, surprisingly, Namgyal had failed to identify himself in the proper Ladakhi manner. "I don't know, I was so surprised by her reaction that nothing came to my mind," he says. "It seems as though she would have doubted anything I'd have said."

We remain silent for a moment, looking at the river. It is here, by the Indus, that decades ago Basgo villagers were executed following Hari Chand's assault on the Hemis Labrang. The turquoise water flows gently, having long ago washed away the traces of an execution that remains forever etched in the memories of a generation of villagers. "I have never been asked these questions in Leh; it doesn't matter much what you look like there," Namgyal reflects. If the habitus Namgyal had developed over the years outside translates into an uneasy fit with village life, he nonetheless was able to craft, over the years, a sense of self while in Leh, where the apprenticeship of village life and skills are not a necessity. But the incident also reveals how fragile this identity is.

"WHEN IT IS WAR, WHAT GOOD CAN HAPPEN?"

Meme Rinchen lives in the upper part of Basgo village, in a house that has seen several generations grow up. Wearing his beard in the traditional Shamma style, a traditional Ladakhi hat, and turquoise earrings, he exudes confidence. *Meme* Rinchen is a former trader and, in the words of his fellow villagers, "has seen a lot of things in his life." The day we meet him, he is sitting outside in the warm sun, calmly spinning his tall prayer wheel.

We chat for some time about his trading years. I eschew the subject of the war. The conversation with the old woman earlier that week, and Namgyal's discomfort with her questions, convinced me that this matter is too sensitive. After some time, I ask *meme* Rinchen's thoughts on environmental change in Ladakh. "There is less water these days," the old man replies. "Look at all the springs (*chu mig*) around, they are all drying up." "Why is it so, *meme-le?*" I ask. *Meme* Rinchen replies with an explanation frequently postulated by Shammas: "It is because there is less snow and rain and because the glacier is almost gone now. When we were young there would be so much snowfall. When we were carrying loads on our backs and traveling, there would be so much snow that we could rest the load on the snow [without taking the load off our back]. You don't get to see snow like that nowadays."

Indeed, it is February and only a thin layer of snow covers the ground. A young woman passing by with her two cows interrupts our conversation, greeting the old man with a large smile before continuing on with the animals. "That's all you can see in the village these days, jersey cows," laments *meme* Rinchen. "Before, people would rear sheep and goats, every house would have one hundred to two hundred sheep and goats. I myself had around a hundred of them. But we don't see these animals anymore." He pauses for a moment, spins his prayer wheel, then continues: "But people have become lazy. They don't want to work, they don't want to take care of animals." Attempting to bring the conversation back to the current water problems, I ask *meme* Rinchen why, in his view, there is less precipitation these days. "I don't really know," he answers. "They (religious leaders) say that people are becoming evil-minded, and so it is because of that." Another pause, another spin of the prayer wheel. "In those days people were good and kind. Nowadays people kill each other. In those days people killing each other was never to be seen or heard. [Long silence.] It is only when the Pakistani war took place that we saw people getting killed. [Long silence.] The Pakistanis had come here and they killed five people in our village,[30] below the palace. [Long silence.] From that time onward, I saw people getting killed."

The long moments of silence that punctuate *meme* Rinchen's reflection signal distress. Bringing these memories back to the surface is obviously difficult. "What happened during the war, *meme-le?* Is this something you would like to share with us?" asks Namgyal. This is one of the many times during my fieldwork that Namgyal forgets that he is "just a [trekking] guide," as he would often say, alluding to his actual means of subsistence. He is so committed and absorbed in the field that he often asks questions, though never in an untactful way, for he has a good sense of the appropriate moment to ask something and the nature of what to ask.

Meme Rinchen continues. He was in his teens when Basgo village was occupied, he explains. It is then that he witnessed "shocking scenes of inhumane behavior." When the massacre at Hemis Labrang took place, he was forced by the raiders to stitch up the torn-apart bodies of the Pakistanis who had fallen during the attack, and to carry the dead to an improvised burial site. "We had no choice, or they would beat us," he recalls. "The bodies were in such a bad state. One of them, a Pakistani officer, was very much damaged . . . his head was missing." He eventually managed to escape from the raiders, putting an end to this gruesome forced work. As he recollects these events, it is clear that decades later the details are still vivid, perhaps more than he would like.

As *meme* Rinchen carries on with his memories of the occupation, explaining how the raiders forced village men to work for them, how homes were looted, and how animals were killed in great numbers, I am unsure whether I should stop him from relating these events from a tragic past. But I decide against it: sharing the experiences seems to bring *meme* Rinchen a certain amount of relief, while Namgyal's role as research assistant morphs into that of confidant. *Meme* Rinchen has apparently forgotten my presence, now addressing Namgyal directly. Ladakh is no exception to the contemporary reality of more people increasingly growing old alone. This loneliness takes on various guises, and a growing generational gap, too, can be a source of isolation. Namgyal's presence has a significant, positive influence on my conversations with elders, who are openly enthusiastic about having a young Ladakhi show interest in their perspective on local history.

While men were subject to mistreatment, detainment, and forced labor, girls often had to run away to hide. "In some instances," *meme* Rinchen says, "some were forcefully used. . . . When it is war, what good can happen?" he asks desolately. In May 1948, when the situation was critically unstable in Sham and the Indian army's promises of reinforcements remained unfulfilled, the president of the Buddhist Association in Leh sent a message to Major-General Thimayya, India's commanding officer in Kashmir, warning that the raiders were now only 27 kilometers from Leh and that "the promises of immediate help had 'deceived' people, otherwise they would have fled and saved at least their lives and the honor of the womenfolk" (Prasad and Pal [1978] 2005:331). Among the testimonies I collected, accounts of rape often emerged, albeit invoked with utmost discretion.[31] One story in particular also marked the people of Sham: one of the Pathan leaders took by force a woman from Saspol, a mother of two children, with him to Baltistan.[32]

Meme Rinchen believes the scenario would have been much different today: "Had they come in these times, they could easily have been killed by all the men of the village." "Why so?" I ask, a bit shocked. "In those days, people were very scared, as killing a human was not known at all. After the war, even the people changed and became bad," he says. Absorbed in thought, he continues: "Maybe because they saw people getting killed and from then, people were no longer scared. They became fearless, but also evil, *konjok*! Nowadays, if 'Pakistan' happens to come, people would kill them, as they have seen lots in their life. Such times have come about." The war, in *meme* Rinchen's view, had transformed the people forever.

Following this forceful assertion, the three of us remain quiet. The silence is occasionally broken by the spinning of *meme* Rinchen's prayer wheel. A cold wind is blowing, perhaps as it had been the winter the raiders departed, leaving the local inhabitants struggling to find a semblance of normalcy. They departed at about at the same time of the year, a bit more than six decades before, but the events are still very much alive in Shammas' memories and have transformed a generation forever. *Meme* Rinchen then shows us his necklace, hidden in his *koncha*. The beautiful object is made of shells removed from his prayer wheel. Placed between the rotating wheel and the handle, he explains, these shells become worn out from the spinning. Each one, of which he has accumulated twenty-three, represents about two years of prayers. The collection, he explains, is the fruit of his prolonged attempt to purge his past.

> I have committed many sins and so I am praying a lot by rotating
> the *mane* (*ma ni lag 'khor*). I have killed many sheep and goats.
> I killed them to eat. In those days, there was a lot of cattle and
> not as many vegetables like today. At that time, it did not occur
> to me that it was a sin to kill them, as I did not know *chos*.[33] As a
> youth, I killed many sheep and goats and ate them as well. I have
> taken many lives, and so I am now doing lots of *mane* to ask for
> forgiveness. I am not sure if I have been able to undo all my sins.
> I think I haven't yet. I have built chortens (*mchord rten*)[34] as well,
> in asking for forgiveness.

A long silence follows as *meme* Rinchen, seemingly lost in thought, gazes at the sky. "I should not have killed the animals," he says softly, "but what can I do? I killed them." Another long pensive silence ensues. After some time, *meme* Rinchen murmurs, "*Om mani ped me om*," as if breathing a long, heavy sigh stifled by the weight of the past.

According to the Tibetan Buddhist conception of life cycle, old age is a time to accumulate merit in view of future reincarnation, whether through prayers or the sponsoring of religious activities. The scene of elders spinning prayer wheels under the sun is a ubiquitous one in Sham. Every elder has their past acts to consider, but the way people judge their own actions remains a deeply intimate affair. Whatever answer he finds in the end, the war was pivotal for *meme* Rinchen, like it was for many other Ladakhis of the older generation. Their views on changes happening today are in part informed by that transformative turning point, when the clash of newly formed states changed the course of local history. As *meme* Rinchen sees it, environmental changes in Ladakh today are inextricably linked to the transformation of people's attitude as a result of the war. His reflection echoes those of many Ladakhi elders who believe that the war, as a critical event, prepared the ground for Ladakhis' fortune to take an unfavorable turn.[35]

A BRIDGE OVER THE INDUS

A short trek westward from Khalatse village leads to a police checkpoint. Here, the road splits. One branch leads further along the Indus toward the Line of Control, while the other heads toward Kargil and Srinagar. Here, movement is strictly monitored, and vehicles must stop at the barrier while an officer checks travelers' papers and inquires about their destination. A few meters from the checkpoint stands a tall chorten, built recently. It faces an iron bridge that spans the Indus. The bridge, like every other in Ladakh, is covered with multiple prayer flags, tattered by the wind. Adjacent to it and slightly lower is a wooden structure: the remnants of another bridge hanging precariously over the river. At one time, the fate of Sham hung in the balance on this very spot (figure 2.2).

On May 22, 1948, during their attack on Khalatse, the raiders were advancing toward the wooden bridge amid heavy fire (Prasad and Pal [1978] 2005:329). At this point, Indian army officials estimated that five hundred raiders were in the Khalatse sector. They were far better equipped with artillery than the defending forces. The mass desertion of the Indian troops following this attack left Khushal Chand with almost no men to resist the invasion. That same night, Khushal Chand made a decision that is remembered in Indian military annals as crucial to saving Leh: he set the bridge on fire before the raiders could cross over the Indus to the northern side, significantly delaying their advance. The wooden bridge was rebuilt shortly thereafter. No longer in service today, its remnants are a fragment of history inscribed in the landscape.

While Khushal Chand and his men were resisting the attack on Khalatse and waiting for reinforcements, fear and suspicion prevailed in the village.

2.2   Wooden bridge over the Indus in Khalatse village, 2013. Rebuilt
       during the first Indo-Pakistani war after it had been destroyed by
       Indian defense and since dismantled.

Outsiders were branded enemies. Salt traders from Chiktan area, who had set up camp on the northern side of the river, and Muslim revenue officials working in the village came under suspicion of spying for the raiders. Wartime justice for crimes real or imagined was carried out quickly: suspected spies were drowned in the Indus. I heard this story on several occasions in Sham, and some of my interlocutors witnessed the tragedy first-hand. There is a striking consensus among the Shammas that this was an extremely regrettable event that should never have happened. Decades after it took place, this incident is still weighted with pending karmic retributions.

According to the law of karma, retribution can manifest itself at any moment. Many Shammas believe that that is what led a relative of Khushal Chand, who is also a reincarnated lama, to sponsor the construction of the chorten that stands close to the bridge where the salt traders lost their lives. Magnificent in its intricacy, its colors still bright, the tall structure contrasts with the surrounding bleak landscape (figure 2.3). Attached to the chorten, dozens of prayer flags flutter in the wind. Behind, the road disappears into the mountains, winding its way toward the Line of Control. The shrine carries two commemorative plaques. One confirms that the chorten, sponsored by the family of Khushal Chand and inaugurated in 2012, honors the contribution that he and his brothers-in-arms made defending the territory in 1948, when Ladakh was "bare and defenseless." The second plaque extols Khushal Chand's courage in more detail:

COLONEL THAKUR KHUSHAL CHAND, MVC
1919–1957
A LIFE OF EXCEPTIONAL COURAGE AND DEDICATION

IEC 9090 Maj Khushal Chand of the 2 Bn The Dogra Regiment was one of the two officers who volunteered to go to Leh in Feb 1948 to help in raising a local militia force and to conduct the defense of Ladakh Valley.

For four months, this young officer, with a platoon of the J&K State Forces and twenty local militia, hastily trained and delayed the enemy advance south towards Leh along the Indus Valley.

Major Khushal Chand conducted a guerilla warfare of a skillful nature, giving the impression to the enemy that he had many more men than he actually had. On one occasion, he held the Khalatse Bridge for 24 hours with just himself and one sepoy. At night, covered by the sepoy, who kept on firing on the opposite river bank, Major Khushal Chand crawled down to the bridge and set it on fire. This delayed the enemy for a further week.

Having no communication with Leh, he made frequent visits there to keep his command in the picture.

Throughout these operations [he was] without proper rations, without mortars, and with an acute shortage of ammunition. Yet Major Khushal Chand led his small band with vigour and skill and by personal example of daring courage. He carried out all his tasks successfully, thus maintaining the high traditions of the Indian Army and setting an example to those serving with him.

The young men of Ladakh whom Major Khushal Chand trained were later transformed into the famous Ladakh Scouts. After the operations in Ladakh, Major Khushal Chand, then barely 29 years of age, was promoted as Lieutenant Colonel. He continued to serve the army with distinction and commanded 9 Dogras. Aged barely 37 years, his brilliant career came to an abrupt end when he died in a plane crash in Apr 1957 in Laos,

2.3   Chorten dedicated to Major Khushal Chand, next to the Indus
       in Khalatse

while serving with the United Nations Organization as part of
the International Control Commission.[36]

Ladakhis believe that building a chorten is a potent way of purifying
negative karma and enables one to gain great merit.[37] There are at least two
stories to every chorten. The first pertains to the identity of the sponsor—an
individual, a family, a group of villagers, an entire community—and is generally common knowledge and easily verified. The second explains the reason
for its construction, which is often the object of speculation. Many people
in Khalatse believe that the chorten facing the bridge was built to purify
negative karma. Inaugurated more than half a century after the deeds it
commemorates, the chorten shows that for the Shammas, the events of the
first Indo-Pakistani war are still very much alive.

CONCLUSION: A CRACK IN THE LANDSCAPE

Not all histories become history; some remain embodied, as in the "arrested
history" of the Chushi Gangdrug resistance in Tibet (McGranahan 2010).

Whether "arrested" for political reasons, or untold because they have been marginalized by a dominant national history, these histories continue to inhabit the people who lived them, sustaining the "pain of belonging"—the need to suppress experience of tragic historical events in order to conform to expected social norms (McGranahan 2010:53). The Partition was traumatic for many Ladakhis. Murder, rape, the killing of animals, and various other abuses perpetrated during the war saddled Ladakhis with painful memories. For a generation of Shammas, memories of these events nurture the pain of belonging, as being Ladakhi means having lived the violence of the war and having been thrown overnight into a context that was, for many, wrought with ethical dilemmas. Between remaining neutral and actively participating in the conflict, there were no easy options for Shammas.

For the Shammas, India's victory came as a great relief and elicited gratitude toward the Indian state, often referred to by elders as "the great Hindustan." Reflecting on the outcome of the war, Ishey Wangail told me: "We suffered a lot under the Pakistanis; we are grateful to Hindustan and Hari Chand"—a reflection I repeatedly encountered among the Shammas of his generation. With the passage of time, the group of Lahaulis morphed into heroic figures in the oral history of Sham. The fact that most of my interlocutors alluded to Hari Chand, whom they described as a *spao* (*dpa bo*), or heroic figure, and even a quasi-supernatural being, is in this respect particularly telling.[38] Moreover, both Hari Chand and Prithi Chand have become part of Ladakhi folklore: when the raiders were expelled, throughout Sham people would sing a song that extols the pair's exploits.[39] Thus, although Ladakhis' contribution was fundamental to the success of the defense operations, in their eyes the initial ragtag army of Lahaulis saved Ladakh.

Shammas' memories might also be nourished by prevailing communal tensions. Many Ladakhis perceive that in the past relations between Buddhists and Muslims were more cordial. A man once told me, reflecting on interactions between communities through trading activities: "In those days there was no differentiation by religion between Balti [Shia Muslims], Khache [Sunni Muslims], or Boto [Buddhists], and the travelers were given whatever they wanted, and they did their own prayers and we did ours. In those days people had good relations with each other." In other words, people were conscious of religious differences, but this did not preclude the existence of congenial relationships (Smith 2013b). However, with the further exacerbation of the politics of religious identity after the war, as Smith (2013a:49) puts it, in Leh District, in tying together religious identity, majority status, and territory, the unresolved India-Pakistan border conflict has charged the landscape "with territorial meanings tied to religion and power."

The first war with Pakistan created a crack in the landscape, a fracture imposed by the attempt to form nations by drawing lines on a map. This fracture is not only territorial; it is also moral. Shammas' memories of the war emphasize the elements that conflicted with their moral values and their cosmology. The crack in the landscape is therefore also about the power of karmic retribution that many Ladakhi elders see in the events of the war. For many, this explains why Ladakh is changing today. The crack is also about the power of the state to reconfigure Ladakh into a border area after the independence of India, with significant demographic and economic consequences. With the decentering of the traditional agro-pastoralist economy, it has become increasingly difficult to live the good life and be a moral person.

CHAPTER THREE

Becoming Sentinel Citizens

The Reconfiguration of Ladakh into a Border Area

Born and brought up in Ladakh, the Nunnus are excellent troops for employment in this region. This is how the Ladakh Scouts came into being. . . . I am honoured and privileged that I am a Commandant of the Ladakh Scouts.

—COLONEL A. P. SINGH, COMMANDANT, LADAKH SCOUTS

Ladakh is a sensitive area situated at two border areas, Pakistan at one hand and China on the other. You require army here. Your existence is in question now. Army and tourists are the twin pillars of Ladakh. We are here for your well-being and your protection. [. . .] People shall understand us. Harmony shall prevail if we appreciate each other.

—COLONEL SONAM WANGCHUK, MAHA VIR CHAKRA, INDIAN
ARMY OFFICER SERVING WITH THE LADAKH SCOUTS

Look at how much development Hindustan has made. Hindustan is very great.

—SONAM MORUP, ELDER OF SKINDEYANG

"ARE WE AT WAR"? I SHOUT HALF-JOKINGLY TO NAMGYAL AND Thinles, still tucked in their *razai*, quilts typical of Pakistan and North India, in the room next to mine. Waking up to the sound of military aircraft is an everyday affair in Leh. The town lies within one of the busiest military air corridors in India. I continue to be amazed that the inhabitants of Leh do not seem to notice this cacophony of noise. What feels deeply intrusive to my ears has become part of their regular soundscape. But this morning, the engine concerto drags on. A few days earlier, Chinese troops made yet another incursion across the disputed Line of Actual Control between India and China.

My friends interpret my question as a signal that my room—the most frequented space in the house, since it is the room with the TV set—is now open for company. Thinles, a friend and schoolteacher who is posted in a village at the border of Pakistan—a place where, he often stresses, "there is not even cellphone network"—enters the kitchen and starts to prepare tea, adding enough condensed milk to induce diabetes. Namgyal joins us and lights a cigarette. "Yesterday I saw many army trucks going toward the border, full of Gurkhas," he says enthusiastically. "They were all impassive, as always. Mercilessly they will kick these Chinese out." A little over a week earlier, on April 15, 2013, about three dozen of China's People's Liberation Army (PLA) soldiers had pitched tents some 20 kilometers inside the Line of Actual Control, in eastern Ladakh.[1] In a further provocative gesture, they unfurled a large red banner on which were inscribed the words, "You have crossed the border, please go back." Yet this was almost business as usual along the border, judging by the announcement a few days later by the Union Ministry of India that there had been more than six hundred incidents of transgression by Chinese troops along India's border with China in the last three years alone—though the vast majority were hidden from the Indian media in order to prevent escalation (Sitaraman 2013). The event of spring 2013, known as the Depsang Incident, in reference to the Depsang Plains where it took place, was more unusual. Tensions were heightened because India had built infrastructure in the vicinity of the disputed zone. Later, in May, the PLA withdrew and the Indian construction activity was halted, both sides having made a show of force to assert their presence.

The first Indo-Pakistani war marked the onset of the reconfiguration of Ladakh into a border area. Now the Line of Control runs through Ladakh, separating the two countries. A bit more than a decade later, the Sino-Indian war (1962) was to further confirm the destiny of Ladakh as a border area. The Chinese People's Liberation Army took control of Aksai Chin, an uninhabited high desert plateau of Ladakh of about 37,000 square kilometers, proving to the Indian state that the borderland had to be better controlled and monitored.[2] The conflict sealed a previously fluid border between Ladakh and Tibet, thus closing centuries-old trade routes. With the conflicts between Pakistan and China, Ladakh, a region that had long been a distant concern for Delhi, became a strategic region and was reconfigured into a border area, a central element in the preservation of the integrity of the Indian nation at its northernmost frontier.

That morning in my room, it is this very border context that makes my friends turn to the question of the role of Ladakhis in the army. Namgyal and Thinles are playing up the skills of Ladakhis in the mountains as a key element in India's victory during the Kargil war. With great passion, my companions recollect accounts of the war that glorify the Ladakhis, who would climb the steep, rocky mountains without equipment; conversely, their fellow soldiers from other parts of India, allegedly less agile in this environment, required ladders, therefore moving at a pace that made them extremely vulnerable to enemy fire. At some point, Namgyal disappears and returns with a picture. Proudly, he points out his uncle, who is serving in the army, posing in the photo with a few other Ladakhi soldiers, assault rifles in hand. Standing imposingly in the foreground, his hand raised in a military salute, is Colonel Sonam Wangchuk, a Ladakhi much celebrated for his exploits during the Kargil war.

During my fieldwork, I regularly heard accounts comparing the skills of Ladakhis against those of their fellow soldiers from the plains of India. These accounts were first-hand, shared both by interlocutors who were serving in the army and by middle-aged retired servicemen. But these stories were never told in a way that intentionally ridiculed the soldiers coming from outside Ladakh. Many of my interlocutors had obvious empathy for those who had come all the way from South India to work in a place they were "not suited for." On one occasion, a close friend of Namgyal's, Sonam, who was serving in the army, came to visit us in Leh, clearly quite distressed. A fellow soldier had died a few days earlier. Having been unable to adapt to the climate of Ladakh, the man from Tamil Nadu had developed a violent form of lung infection and passed away. Sonam was in shock, wondering why men from the south should be sent to the mountains in the first place. Sonam's attitude is not uncommon among Ladakhis: instead of calling into question the very geopolitical context of Ladakh, many Ladakhis seem to have reimagined themselves as the most apt people to protect the border, precisely because this sensitive terrain is their land.

How is it that the Indian army is not seen as an occupying force in Ladakh? Why is the state's territorialization strategy in this part of the country, largely based on militarization, by and large accepted by the local population, despite all the disrupting changes it brings to the region?[3] In this respect, the situation of Ladakh differs significantly from that of other parts of Jammu and Kashmir state.[4] If, as Namgyal and Thinles's attitude illustrates, stories of Ladakhis at war are infused with the ideals of manhood and militarized masculinities noted by Aggarwal (2004:219) and Bhan (2009:143), they also reveal the affective nature of military labor in Ladakh. The context of the onset of the formation of the Indian state in Ladakh has nurtured among a generation of Ladakhis ideas of both the state and its military agents as liberators.[5] The discourse about "the great Hindustan" they hold is, however, one that is unique to their generation. In contrast, today many Ladakhis have internalized a responsibility for the integrity of the nation's sovereignty. To be sure, not all Ladakhis are satisfied by the presence of the army, but their concerns tend to be oriented toward land grabbing by the Indian army. By and large, it has become an accepted fact that Ladakh is a militarized region, something Ladakhis rarely question.[6]

Ladakhis' sense of responsibility toward protecting their land and their effective role in defending the territory must be understood in light of the affective transactions between Ladakhis and the state, which inform state production in the region. The affective basis of statemaking in Ladakh is a two-way mechanism. First, the events of the first Indo-Pakistani war, as they unfolded in Sham, were determinants for the history of postcolonial India. As much as the formation of the Indian state took place in a context of war into which Ladakhis were thrust suddenly and unexpectedly, the war also thrust military agents of the state into a region with which they were unfamiliar. The notion of "sympathy" developed by anthropologist Danilyn Rutherford (2009) allows us to consider the sentiment among military agents endeavoring in Ladakh during the first Indo-Pakistani war. The historical conjuncture that led to the formation of the Indian state in Ladakh, together with the materiality of the region's environment, led to the emergence of affective feelings for a local population.

These feelings also have implications for the role played by Ladakhis in the production of the state in the region and remain, since the events of the Indo-Pakistani war, central to contemporary imaginations about the Himalayan frontier of India. Military work in Ladakh is therefore an affective labor, namely work that is motivated by and leads to the emergence of sentiments. Anthropologist Ravina Aggarwal (2004) examines how, in Ladakh, nationalism is questioned, nurtured, and reinforced through cultural performances and state-sponsored festivals, always in a way that causes Ladakhis to see themselves as central to the nation's project. Along the same lines, anthropologist Mona Bhan (2008) examines how national consciousness is fostered among the Brogpa community[7] of Ladakh through the notion of *khral*, or mandatory work (akin to *begar* work), which since the independence of India has come to encompass work at the border for the military in time of crisis. Moreover, Aggarwal and Bhan (2009) study how the Indian

3.1 War memorial at the Hall of Fame Museum in Leh, commemorating
military personnel who lost their lives in Ladakh

army strategically institutionalizes compassion in social development pro-
grams as a means to secure the loyalty of populations living in zones contiguous with the border with Pakistan in Ladakh. Building on this work
allows us to see how affect and feelings of care are central to the production
of the Indian state in the region as Ladakhis cultivate sentiments of responsibility through their endeavors to protect the land. This work is not only
individual, but also collective, for even if not all Ladakhis work in the army,
the vast majority are aware of the significant role played by their community in the protection of the border and a large number take pride in this
involvement.

The affective terms of local geopolitical processes produce a "border subjectivity" (Aggarwal 2004:3), which takes shape through labor practices that
include not only direct military work but also a subsistence contingent on
the very presence of the army, both as a catalyzer of a new household economy and as a custodian of geopolitical stability. This is made possible by
the sentiments of care and sacrifice for the land enabled by the borderland
subjectivities. Ultimately, the Indian state has contributed to the production
of a "sentinel citizenship," a mode of belonging to Ladakh that is enabled by
the material reconfiguration of the region into a border area, the mobilization of the local population for military labor, and an affectively structured
relationship between Ladakhis and the state. Sentinel citizenship is also a

moralized form of being a citizen, which entails a form of care, manifest in a feeling of responsibility toward the land.

STIRRING UP A SYMPATHETIC IDENTIFICATION

Stretching about 2,400 kilometers across the north of the subcontinent, the Himalayas ("abode of snow" in Sanskrit) have historically been viewed as a natural frontier. The oldest known source describing the mountain range in this manner is the *Vishnu Purana* (about the first century BCE), but this is also a trope that permeates British accounts from the colonial era (N. Mathur 2013:72). Aggarwal and Bhan (2009:521) note that the capture of Ladakh by the Dogra rulers of Jammu was a crucial historical moment in the construction of Ladakh as a "natural" border for India.[8] These ideas are well reflected in British army engineer Alexander Cunningham's *Ladakh: Physical, Statistical, and Historical* ([1854] 2005:41–81), which depicts the different mountain chains surrounding Ladakh as natural boundaries with neighboring regions and countries.

Under the British colonial administration, it was never deemed important to have a heavy physical presence of the state and its institutional apparatus in the region. The fact that the mountains of Ladakh and its surroundings are arduous to traverse certainly contributed to the idea that substantial state presence on the ground was unnecessary, something that would change dramatically after the independence of India. Descriptions of Ladakh as a harsh, dangerous, and unexplored hinterland pervade British colonial accounts as well as those of the various European missionaries who traveled to Ladakh (Jina 1994). For instance, while the British explorer and army officer Francis Younghusband was eloquent about his passion for the mountains and his taste for adventure and often depicted himself as a braver of difficult terrains—frequently in self-aggrandizing language—he was nevertheless quite explicit about his fear when crossing hazardous passes. This is notably the case when he crossed the Khardung pass in Ladakh, an ascension during which he experienced severe altitude sickness (Younghusband [1904] 2009:114). Moreover, observations on the harshness of Ladakh and the sense of desolation felt in the face of what is often described as an arid and barren landscape abound in these accounts (see Aggarwal 2004:114–16).[9] Thus, if, as Peter Bishop (1989) argues, Western travelers have often described the Himalayas as an archetypal utopia filled with mythical fantasies of Victorian romanticism, the same literature suggests that Ladakh often conveyed an opposite feeling.

The defense of Ladakh during the first Indo-Pakistani war proved difficult. Besides the lack of equipment and military personnel, the very terrain on which this war was taking place proved to be a challenge. For one, Ladakh was largely unknown to the Indian army. An Indian army colonel who fought in Ladakh during the war told Virendra Verma (1998:15), former principal of the National Defence Academy, that "when his Company was ordered to proceed to Leh via Manali in 1948, its officers felt so baffled, as they were totally ignorant of the geography of Leh-Ladakh. They had to rush to the libraries to consult books and atlases to find out more details about these parts of India. Never before had the people of India, in general, been in contact with these regions, nor had the Indian Army units during British times ever landed there." The struggle of military agents to protect the sovereignty of India during the war was also a struggle to survive and operate in an inhospitable and largely unknown terrain. This context of military warfare and unfamiliarity with the land became conducive to the emergence of affective feelings for a local population.

Sentiments, scholars have argued, may be socially determined, that is, the product of certain ideologies and institutional structures (Foucault 1976; Povinelli 2006). The relation between affective sentiments and the state has been the object of recent scrutiny (Muehlebach 2011; Navaro-Yashin 2012), with scholars demonstrating how empathy, sympathy, and compassion, feelings that encompass a range of "emotional vicarious attitudes" (Bubandt 2009:566), are produced as a result of interactions between citizens and the state. Analyzing colonial archives, scholars have identified the historical contingency of these sentiments and forcefully demonstrated their significance to political projects (Rutherford 2009; Stoler 2004).

In her fascinating study of Dutch colonialism in New Guinea, Danilyn Rutherford suggests that both the assessment and fostering of affective relations and sentiments between the state and its citizens was a defining feature of colonial expansion. Building on the empirical philosophy of David Hume, she develops a materialist conception of sympathy, a term that, as she defines it, "does some of the same work as terms like recognition or identification." As an empirical concept, sympathy "tracks the intricate pathways through which encounters with objects and others give rise to feelings and thoughts." Sympathy-as-affect is here a "process grounded in embodied interaction" and therefore intricately linked to the materiality of encounters (Rutherford 2009:4–8).

For agents of the Dutch empire, Rutherford maintains, sympathy was not only a feeling through which they could relate to a local population; it also had practical implications. Through sympathy, agents sought to expand their influence by "shaping the inference" of their colonial subjects and therefore shaping their sentiments and their inclinations. In this respect, sympathy is also fundamentally about imagination. From distant Europe, New Guinea was imagined as belonging to the Stone Age era, a belief that agents of the empire were well exposed to before their departure for the colony. This work of imagination was indeed an artifice that enabled the extension of sympathy: in order to shape the sentiments of their colonial subjects, officials realized they could not think through their own institutions and technologies. Rather, they had to enter the Papuan's world and "build an imagined community of empire out of lived experience, one interaction at a time" (Rutherford 2009:6).

Rutherford's analysis is conducive to understanding how imagination and embodied interaction in an unfamiliar environment are also at work in the emergence of the state's affective feelings for Ladakhis in a context of military imperatives. Ladakh had long been imagined as a distant forbidden place where the environment is inhospitable to humans, and this is where Indian military agents found themselves during the first Indo-Pakistani war. As we saw in the previous chapter, the Indian army strongly relied on Ladakhis' support during the war. In his account of the war, Prithi Chand makes it clear that Ladakhis' assistance, even when unarmed, proved crucial to the Indian army's success. For instance, Prithi Chand reports how during one visit in the Nubra Valley he and his men suffered from altitude sickness, a state he overcame by using a local remedy—locally made alcohol. To treat their injuries, soldiers relied on *amchi* (*am chi*), the local practitioners of traditional Tibetan medicine (Chibber 1998:155–56). When rice supplies were exhausted in Ladakh, soldiers eventually came to enjoy the local barley meal, which proved much more suited to the cold weather (Chibber 1998:183–84). What these examples show is that for the Indian soldiers, the experience of the war was as much about coming to terms with an unknown place and an unfamiliar environment as about fighting the enemy. These men had to immerse themselves in a different world, and the only way to survive it amid the circumstances of the war was to rely on the local population.

Vicarious feelings, it has been observed, may arise at critical historical moments, amid unsettling and unpredictable circumstances (Berlant 2004:7). As such, the embodied experience of the mountains of Ladakh and the war provided the context for affect-laden encounters. The defense of the territory therefore came to be predicated on relatedness between these military agents and the local population, as Indian soldiers were acutely aware of their dependence on the local population. But these interactions and sentiments were to be consequential. Rutherford argues that although sympathy is a feeling that may emerge spontaneously during interactions, it also shapes ensuing interactions as it "registers a gradient of feeling and moral obligation, a sense of two subject positions and the interests associated with them being more or less tightly aligned" (Rutherford 2009:5). The sympathy that infused the relationship between military agents and the local population in Ladakh was also consequential. Driven by military considerations, the state soon realized the crucial role the local population could play in defending the vast, inhospitable, and uncharted territory of Ladakh. Thus, if the experience of the war stirred a sympathetic identification, these same feelings would become a defining feature of statecraft in Ladakh, for it enabled the Indian state to reimagine its Himalayan frontier.[10]

REIMAGINING THE HIMALAYAN FRONTIER AND ITS PEOPLE

If ideas about Ladakh as a forbidden hinterland and about the Himalayas as a natural boundary for the subcontinent contributed to a weak presence of the colonial state in Ladakh during the British administration in India, the perception of its inhabitants as harmless people must also be taken into consideration. In fact, in terms of security, Ladakh as a region has never been a significant concern for the British, who were more preoccupied by the threat coming from Gilgit (van Beek 1996:89). Ladakhis were seen as harmless, cheerful, and, as we saw in chapter 1, as lacking courage in the face of abusive authority.[11] These depictions contrast markedly with the idioms of fierceness and barbarianism in British colonial accounts of neighboring frontier communities, such as the people of Hunza in Baltistan (see Hussain 2015). In other words, Ladakhis were seen as harmless, causing no threat to the British Empire at its northern frontier. This is also reflected in the frontier governmentality of the imperial administration: unlike their neighbors the Pathans, Ladakhis were excluded from the draconian Frontier Crimes Regulation, an administrative measure and legal code developed by the British Raj to govern the problematic "wild tribes" of the hills along the northwest frontier of the empire (Hopkins 2015).

After the wars with Pakistan and China, the long-assumed "naturalness" of the Himalayas as a border was far from an immutable fact. Ladakh could no longer be a distant concern for Delhi but, rather, had to be transformed into a mountain fortress for the subcontinent. Asserting control over the territory of Ladakh became a priority for the Indian state following Independence. The topography of Ladakh plays a significant role in the very unique statecraft by which Ladakhis are mobilized to protect the Himalayan frontier of India. As Bruno Latour (2004) argues, the materiality of nature is central to its political configuration. Yet the prevailing narrative of human domination over nature, so often depicted as entirely tamable by human hands, is severely challenged in the Himalayas. In Ladakh, the governing of a landscape made up of high peaks and steep slopes, with desert conditions, freezing temperatures, and high altitudes, challenges human endurance, defies technological capabilities, and often eludes expert knowledge.

The institutionalization of Ladakhis' aptitude for mountain warfare, as revealed amid the geopolitical circumstances that marked the onset of the formation of the Indian state in Ladakh, has been one of India's key strategies. But this process also entailed the reimagination of this local population. The innocent Ladakhis of the colonial era were now seen as fierce warriors of the mountains, and their resilience became a key asset to the production of the state in this mountain area.[12] In particular, the ability of Ladakhis to navigate the mountains and their seemingly unwavering positive attitude in the face of the harshness of military life in this terrain now constituted a fundamental asset to the defense of the northernmost frontier of India. Reflecting on the work of Ladakhis in the Nubra Valley during the first Indo-Pakistani war, Virendra Verma (1998:51–52) reports:

> The Nubrans were the sons of the soil. They could climb steep slopes of the mountains with perfect ease and descend or roll down fast without any difficulty. They became known for their mobility; agility and nimbleness was second nature with them. The Nunnus [*nonos*] would always be present where they were least expected. They would climb up the snow-covered heights ranging from five to six thousand meters and they would then roll down to come as close as possible to the bunkers of the enemy forces positioned about a hundred meters below to hurl hand grenades. . . . Unmindful of inclement weather, biting cold winds or snow blizzards . . . Nunnus could battle their way forward through waist-high snow. . . . They never complain of frost-bite.

Verma concludes that these physical aptitudes were crucial in determining the outcome of the first Indo-Pakistani war. Similarly, a recent article published in *Asia Defence News* praises the endurance of the *nonos* for "operating efficiently in extreme hazardous terrain and inhuman weather conditions," adding that "these brave sons of the soil have set standards and benchmarks for others to emulate" (Renu 2013:31).

These depictions—arguably essentializing—show that the military state sees the local population as instrumental to the process of territoriality.[13] They also reveal that although Ladakh may still be depicted as a harsh and hostile environment, its population was reimagined after the independence of India. The passive Ladakhi of British colonial accounts, fully submissive in the face of the abusive rule of the Dogra and displaying no bravery in the mountains, was now described as fierce and resilient, helping to build the nation in a highly unforgiving and contested terrain. These ideas would form the modalities of the reconfiguration of Ladakh into a border area, that is, through the active involvement of the local population.

Assembling the Ladakh Scouts after the Sino-Indian war in order to institutionalize Ladakhis' aptitude for mountain warfare was a crucial step in rethinking the Himalayan frontier of India and its local population.[14] Besides the Scouts, contingents from all over India are based in the region, and it is through both this military structure and embodied interactions in a harsh and unpredictable environment that the military continues to develop sympathy for a local population. This is tangible in testimonies from fellow military men who are called to serve in Ladakh. A document entitled "Nunnu—An Ideal Soldier" discusses the notes of Colonel A. P. Singh during his time as commandant of the Ladakh Scouts (Verma 1998:136–38). Singh explains that, being "eager to learn as much as he could about 'the Nunnus,'" he asked his non-Ladakhi officers to describe their experience among Ladakhi infantry.[15] The testimonies he collected reveal the great esteem officers had for the fortitude, resourcefulness, and work ethic of the Ladakhis.

One officer, for example, reported how a commanding officer got lost during a nighttime training exercise above 5,000 meters. He praised the *nonos* for volunteering to search for him through the night, even though he was not a Ladakh Scout. Another officer reported how, when the communication line was cut at a post located at an altitude of 6,100 meters, two *nonos* were asked to go fix the problem the next morning, but instead went on their own at night and returned only seventeen hours later, having completed the task with "nothing to eat except the snow" Another officer explained, "I have no reservation in saying that the Nunnus are undoubtedly par excellence," adding that they are "simple, hardy, brave, honest, trustworthy, dependable, highly co-operative, resourceful, enterprising and always happy" (Verma 1998:138). If the trope of happiness still infuses depictions of Ladakhis, it is now a happiness that is accompanied by an infinite ability to endure military operations in the mountains, an instrumental asset for the Indian state in defending its territorial sovereignty on the Himalayan frontier.

It is also through military work in the mountains that intimate identification with Ladakhis takes place. Prior to their departure for Ladakh, officers are well informed about the hardships of the terrain they will be operating in. It is well known that every year a number of military men coming from the plains of India lose their life in the mountains, not under artillery fire, but rather due to acute mountain sickness. Moreover, there is always the risk of an avalanche or of falling into a crevasse. But military work in the mountains and interactions with Ladakhis causes these officers to identify with Ladakhis. One officer reported: "On the very first day I landed in the company, I could make out that I was totally acceptable to the Nunnus and I am sure each and every officer who came to the Ladakh Scouts feels the same" (Verma 1998:138). In the broader context of India, Ladakhis are considered to be part of tribal populations, groups that are generally marginalized. If the "Other" has often been considered with suspicion, what this officer's comment reveals is that the geopolitical context of postcolonial India and the work required to secure the border in a difficult terrain leads to an unlikely intimate identification.

The way Ladakhis have been set up by the Indian state as warriors of the mountains has much resonance with how the British set up mountain peoples as a martial race during the colonial era, as is the case with the Gurkha of Nepal. In his analysis of writings by British military officers about the Gurkha, anthropologist Lionel Caplan argues that if these depictions are orientalizing, in the sense that they function as elements of neocolonial domination, they are, however, not "a species of orientalism whose overriding concern is domination, or control." Rather, these depictions enable a process of self-identification, by which "British officers are engaged in a conversation with themselves, the Gurkha serving primarily as a means of self-reflection" (Caplan 1991:594). In a similar way, the characterization of Ladakhi soldiers as a resilient people possessing the perfect skills to protect the nation at its northernmost frontier has become a means through which the military state reinvents the Himalayas as a border and, ultimately, rethinks its sovereignty.

Following the Sino-Indian War, the Scouts played a significant role in subsequent wars with Pakistan in 1965, 1971, and 1999. The extensive media coverage of the Kargil war of 1999 greatly contributed to extant sentiments of sympathy toward Ladakhis in the public sphere. Stories depicting the aptitude of Ladakh Scouts for high-altitude warfare and the determining role they played in the outcome of the war pervaded Indian television channels and print media at the time.[16] The public affection for the Ladakh Scouts since the Kargil war is obvious in various media publications, blog posts, editorial comments, and popular websites that praise their hard work. The monikers the Ladakh Scouts have earned over the years are telling: they are sometimes referred to as "snow tigers" and "snow warriors" and are also said to be "eyes and ears" of the Indian army.

Through affective transactions between Ladakhis and the state, which continue to be central to statemaking in the region, a sentinel citizenship has been fostered by the new border context. Close to seven decades after the Partition, the conflict with Pakistan remains unresolved, and the faith of Ladakh as a border area has been further confirmed by unreconciled tensions with China. The entry to Indian nationhood in Ladakh has thus been one by which, for Ladakhis, the fabric of everyday life has become impacted by a new geopolitical context, whether one directly labors for the army or not. Today, the vast majority of Ladakhis have at least one family member working for the armed forces, opportunities that bring men outside the villages for most of the year (see also Ladags Melong 2005:29; Tsewang Rigzin 2005).[17] Besides enrolling in the Ladakh Scouts, many Ladakhis fill various logistical support functions for the army, ranging from porterage to skilled technical services. According to some estimates, outside of military service per se, about a third of Ladakhi men earn their income from the military, through different types of work (Angeles and Tarbotton 2001:102).[18] Many are also employed in the various branches of the state's paramilitary apparatus, such as the Indo-Tibetan Border Police and the Sashastra Seema Bal of the Central Armed Police Forces. As Fewkes (2009:24) points out, given the significance of the army in contemporary Ladakhi economy and the fact that Ladakhis are an integral part of the Indian army, we should "not assume a complete separation between the interests of a 'native' population and the Indian army." With the ever-growing military presence in the social and economic life of Ladakh, the border context has become a pervasive one in the region, and this has significant implications for life in the villages.

Yet as Rutherford (2009:6–7) notes, sympathy can be "partial," and accordingly, the affective military state is also a selective state. For one thing, scholars have demonstrated the antagonism of the Indian state to the Muslim communities in the border area of Jammu and Kashmir, the state that governs the administration of Ladakh (Jones 2009; S. Mathur 2016).[19] Moreover, given that relations of mistrust became the basis for the many post–Kargil war military interventions in the region and that Muslims living along the border in Jammu and Kashmir state have historically not been granted the same roles and responsibilities as Buddhist Ladakhis with respect to the military production of the state, it is fair to say that the feelings of the Indian state might also be driven by communalism.[20] The public thus produced by sentinel citizenship, it must be noted, remains only a fragmented one.[21] But Ladakhis' contribution to the reconfiguration of the region into a border area should not be considered as labor in solely economic terms. The values of morality and loyalty toward the land and notions of collective sacrifice that infuse statemaking are equally crucial to fostering a sentinel citizenship.

THE PERVASIVENESS OF THE BORDER

Back in my room in Leh, while we are discussing matters of warfare in Ladakh, the electricity suddenly comes on, bringing to life the popular TV set, my computer, a modem, and a bare light bulb. The TV channels are buzzing with news of the PLA's intrusion beyond the Line of Actual Control. The news broadcast airs a perpetual loop of three images: a map with the location of Daulat Beg Oldi, an Indian military camp about 30 kilometers from the incident; some military bunkers; and the seemingly eternal landing of an Indian Super Hercules military aircraft against the backdrop of a lunar Ladakhi landscape. TV commentators pour forth alarmist pronouncements on territorial revisionism and Chinese intentions. There is a news report that the head of India's army will soon visit Jammu to assess the security situation. On my computer, I find a *New York Times* article reporting on the incident: news of the situation has reached the international media.

"Stupid them," exclaims Namgyal reproachfully. "Stupid media, why are they showing these images? They are going to ruin the tourist season." It is midspring, and the scramble for revenues from the tourist industry is about to start. Despite a very short season—from May to August—tourism brings a significant monetary influx to Ladakh. Namgyal fears that these images will discourage tourists from coming. This is no time for border skirmishes: many people have to pay off debts accumulated during the winter. The television suddenly shuts down as the power goes off once more. While elected politicians are quick to boast of the growing electrification of Ladakh as a political achievement, they do not mention that electricity outages are the norm and having power is the exception rather than the rule. "I think that, if given the choice, many Ladakhis would prefer to be under China rather than India," suggests Namgyal, visibly annoyed. "At least they have good roads and power there."

Namgyal's remark is more the product of cynicism than of dissidence. In fact, the patriot sentinel has become for Ladakhis an important self-image (see van Beek 1998a; Vasan 2017:48). His discontent simply expresses the practical realities of life: subsistence, work, debt payments, and a yearning for a stable environment. Many others shared Namgyal's concerns about the media's depictions of the Chinese incursion. Several tour operators even appealed to the minister of tourism of Jammu and Kashmir state, believing he could intervene by reassuring the public about the situation in the region. The blame was placed entirely on the media, and none of my interlocutors mentioned Ladakh's status as a border area amid enduring geopolitical conflicts. For many, the immediate concern was the potential threat the Chinese incursion could have on the regional economy, rather than its potential for violence. This situation can only be explained by the fact that these incursions have become close to commonplace for Ladakhis. In fact, while episodes of this kind may remain unknown to the general public, for Ladakhis, many of whom are employed in the army or have family members who are, they are unexceptional. But the media intervention, in exposing the situation of Ladakh, served as a painful reminder to Ladakhis of the volatility of what they rationalize as normal life. The Depsang Incident also contributes to underlining the pervasiveness of the borders separating Ladakh from neighboring China and Pakistan. These contested boundaries are potential grounds for war that affect an entire region.

Global contemporary processes, ranging from travel, capitalist flows, and the circulation of media, as scholars have demonstrated, are generative of new forms of subjectivity and identity (Appadurai 1996; Ong 1999). But a new sense of self may also emerge as transformative processes, such as geopolitical ones, reconfigure a place and its meaning. In Ladakh, the transformative flow of material and ideas is military in nature, and the regular border skirmishes and intrusions constantly remind Ladakhis that the mundane is taking place against the backdrop of a sensitive borderland. What the Depsang Incident also demonstrates is that the border context is woven into the fabric of everyday life: with the independence of India, Ladakhis have become citizens at the heart of regional geopolitical conflicts, a situation that has contributed to transforming the region politically, economically and socially.

• • •

It is a beautiful sunny day. On the way to the Ladakh Scouts' base near Phyang village, we pass in front of a military installation. The insignia at the entrance to the site carries the words *Kiki So So Lhargyalo*—"Victory to the mountain gods"—the motto of the Ladakh Scouts. Stein (1972:206) interprets the invocation as a war cry, reflecting "the warlike nature of the gods and the idea of passing through a dangerous and strategic place." A nonagenarian in Nyemo once told me that decades ago, during the first war with Pakistan, villagers realized that the Pakistanis had run away only when they heard Ladakhis shouting "Kiki So So Lhargyalo." Traditionally, Ladakhis would shout this phrase when reaching a mountain pass as a rite of respect and deference to the local mountain gods.[22] The invocation of the motto during the war illustrates how, for some Ladakhis, the dispute not only is taking place in the realm of humans, but also involves the deities, through the support they provide. This cultural interpretation, in becoming an emblem of the Ladakh Scouts, further contributes to legitimizing the role of Buddhist Ladakhis in the military production of the state, as their endeavor receives the benevolence of the deities.

We are on our way to the Ladakh Scouts' Golden Jubilee, celebrating the fifty years of the regiment. In the vehicle, along with myself and Namgyal, are Thinles and his brother, a monk appointed to a remote village and who has come to Leh for the event, a journey that took ten hours by bus. I accepted the invitation reluctantly, since the conditions of my research visa forbid my entry into any military sites. But my friends insist that this is a special event, "a celebration with free food and games," where everybody is welcome. Somewhere along the long, narrow road leading to the celebration, two soldiers stop us for routine verification. I apprehensively hand my passport to one of them, explaining the note on my visa. My gesture elicits no more than an exchange between the man and Thinles on the triviality of my concerns; a moment later, the soldier hands me my passport and Thinles drives on.[23] Arriving at the site of the celebrations, we were greeted by the Ladakh Scouts' insignia: a standing ibex with head proudly raised.[24] Along the way, we stop for a group picture under a colorful arch proclaiming the site as the "highest training center in the world." In a mixture of religious and military graphic elements, the inscription is bounded by the eight auspicious symbols of Buddhism (Ashtamangala).[25]

We finally reach the ceremonial grounds, consisting of an open court surrounded by colorful tents, where food is being served. The place teems with civilians, infantrymen, a few women, children, and many monks. The day before, high-profile army officers took the stage to address new Ladakh Scout recruits who were brought here with their parents and relatives to share a ceremonial cake. On one side of the grounds, children and young men form a large circle to play a ball game under the direction of an elderly woman. Winners are awarded small gifts. We sit, eating and chatting, for some time. Namgyal, Thinles, and his brother bump into old friends they have not seen in years. The ambience is jovial and welcoming. The religious intermingle with the military, families celebrate against the backdrop of artillery. The activities and the atmosphere are quite evocative of the popular winter monastic festivals, where people from all parts of Ladakh gather to participate in ritual dances and other ceremonies. These festivals are not only pilgrimages for the most devoted, but also an occasion to socialize with friends and relatives, conduct business in the makeshift markets, play games, and even indulge in gambling or courting.

A public display of affection by the Indian state, the Golden Jubilee of the Ladakh Scouts also exemplifies how the military production of the state has become a central feature of belonging in Ladakh today. The army occupies a prominent role in the local household economy, and the opportunity of employment it provides constitutes a source of aspiration for a younger generation of Ladakhis. Yet the economic incentive to join the army, undeniably strong, does not preclude the existence of the moral feelings attached to this work. Ladakhis who are intimately affected by every spark of tension at the borders because they or a close relative are serving in the army explain, "We have no choice: this is our work" or "this is our duty."[26] Thus, it is not uncommon to find Ladakhis using the language of morality to speak of the collective sacrifice that is necessary for the protection of their land.[27]

Mona Bhan (2008; 2014: chapter 3) notes the normalization of military duty among the Ladakh Brogpa community, which lives close to the India-Pakistan border. Bhan maintains that this work is integrated into the Brogpas' daily routines by being rationalized as community work. It also contributes to fostering a sense of nationalism and reconfiguring aspirations. These insights can be extrapolated at the regional level by drawing from the scholarship on affective labor. Affective labor, or what Michael Hardt (1999:89) describes as "the processes whereby our laboring practices produce collective subjectivities, produce sociality, and ultimately produce society itself," is often based on the commodification of feelings of care and morality and is double-edged in nature (Buch 2013; Constable 2009; Faier 2009).

In her 2012 book *The Moral Neoliberal*, anthropologist Andrea Muehlebach examines how the fetishizing of volunteer work as a means to provide care to destitute and disadvantaged people and groups neglected by the state constitutes a transformative experience by which citizenship, for both those who provide care and those they help, is being fostered as a virtue rather than constituting a positive right. Largely mediated by the Italian state, this labor regime is transforming a population, in particular the unemployed, youth, and retirees, into "affectively laboring citizens" by nurturing the idea that a society needs a distribution of social compassion. This type of ethical citizenship constitutes an affective state that calls on a sense of morality and responsibility. In Italy, this new labor regime found its purchase out of "already available religious-cultural meanings and practices," namely the Catholic Church's long history of "massive solicitation of sacrificial feeling and action in the name of the public good" (Muehlebach 2011:60–61).

In a similar vein, the sentinel citizenship fostered through military labor in Ladakh calls on values of morality and loyalty toward the land. These values were part of the military discourse during the first Indo-Pakistani war and continue to inform this discourse today. The reflections by Ladakhi colonel Sonam Wangchuk in the opening of this chapter, shared in the context of an interview on the presence of the army in the region and the participation of Ladakhis in this work, epitomizes how sentiments of loyalty and responsibility toward the protection of the land are cultivated in public discourses. The army is largely depicted as an ineluctable presence, a necessity that calls for a collective sacrifice, either through direct labor or by accepting to coexist with this military presence, with all the implications that may entail.

Such discourse fosters an affect-laden citizenship, a form of belonging that is cultivated through the values of care and sacrifice associated with defending the land and institutionalized in the borderland context of postcolonial India through military labor. This form of belonging is also achieved through everyday practices, through which the presence and role of the army plays an integral part in the life of Ladakhis. As events like the Depsang Incident and the Golden Jubilee demonstrate, the military in Ladakh calls on a collective sentiment, is imbricated into the intimacy of the family, and is legitimized by an existence that may at any time be compromised due to regional geopolitical conflicts.

◆　◆　◆

But sentinel citizenship neither exists in isolation from the traditional agropastoralist economy, nor does it erase the values with which this work is invested, as a situation faced by Namgyal that spring illustrates. Namgyal's father is a contractor working for the army, and, having been assigned to a project in a remote area, he could not come back to his village to plow his land. Namgyal brought this situation to my attention by informing me that he might have to take a few days off from his work with me. The fact that his father had been absent for pretty much all of Namgyal's life does not exempt Namgyal from this responsibility, nor can he disregard this request, but he nonetheless alludes to the irony of the situation. "When it comes to this, there is no choice; we need to help, no matter the shape or the color of your family," he explains in a cynical tone. The new military economy does not mitigate the abiding moral imperative to support family members through agrarian work and the necessity to farm the land for the household economy; rather, it complicates things, as it generates new dilemmas. This constitutes the Janus-face of the governance over the landscape of Ladakh, namely the displacement of the agro-pastoralist economy. The dilemma expressed by Namgyal over possibly having to support his father with farm work offers a glimpse into what a growing number of Ladakhis are confronted with today, namely juggling work-related responsibilities with the need to take care of the family's land and animals.

CONCLUSION: THE SEDIMENTED BORDER AND BEYOND

Historically a peripheral space, the region of Ladakh was integrated into the Indian state after Independence on the basis of an affectively structured relationship between the state and the local population. The historical conjuncture that led to the formation of the Indian state in Ladakh during the first Indo-Pakistani war, together with the materiality of the region's environment, led to the emergence of affective feelings for a local population as Indian troops found themselves having to come to terms with an unknown and difficult terrain. The defense of the region came to be predicated on the co-optation of the Ladakhis' knowledge of the territory and their ability to navigate their native high-altitude Himalayan landscape. Since then, Ladakhis have become a key workforce for the army, providing an effective labor force for the militarization of the region. Their mountain skills have become institutionalized in the military apparatus, something that has become central to the project of state formation in the region, which is oriented toward the reconfiguration of Ladakh into a border area.

Hence, if for a generation of Ladakhis the war has led to the rise of affective nationalism, developed through the encounter with the state, which in its most laudable form was incarnated by charismatic military figures who changed the destiny of a region, it also led to the rise of sympathy on the part of the state for a local population. As scholars have argued, imaginations about nature and justification for state formation projects often intersect (Cederlöf and Sivaramakrishnan 2006). In Ladakh, the historical conjuncture of the formation of the Indian nation with the materiality of the environment has thus caused the state to reimagine its Himalayan frontier and its inhabitants. Ladakhis' resilience in the face of hardship became a key asset in the Indian state's assertion of territorial sovereignty at its Himalayan frontier. The fatalistic and soft Ladakhis described in colonial accounts suddenly became, in the eyes of the military state, warriors of the mountains. While wars have changed the conception of the Himalayas as a natural border for India, imaginaries about Ladakhis have enabled the rethinking of territorial sovereignty at the Himalayan frontier of India.

Feelings and sentiments continue to be a dominant feature of statecraft in Ladakh today, and they are fundamental to the shaping of a borderland subjectivity. The material reconfiguration of Ladakh into a border area, in involving Ladakhis in the defense of the territory, whether as direct labor or through the sacrifice it entails for a whole community, has produced a sentinel citizenship. The geopolitical context of Ladakh has led many Ladakhis to extend their moral imagination toward protecting their land and to reimagine themselves as having a unique physical capacity to contribute to this project. The prevailing wisdom that the militarization of the region is an ineluctable condition remains largely unchallenged by Ladakhis. Hence, if sentinel citizenship is anchored into a political project, a politics of nature by which the mountains of Ladakh must be a fortress for India, it is also, for Ladakhis, "citizenship as something that arises from the heart" (Muehlebach 2012:51; see also Stoler 2002).

While the ethical responsibility toward the land has traditionally been anchored in agrarian and pastoralist work, the geopolitical context of Ladakh today has allowed for this anchoring to diversify and include military work. But Ladakhis' participation in this economy makes it increasingly difficult to maintain traditional subsistence activities. As the border context sediments, both through an expanding military presence and because the border context permeates different spheres of life, the political and environmental processes that transform the landscape of Ladakh today morph into the local reality and become incorporated into the narratives about this place. This is particularly true for an aging generation of Ladakhis, one that lives with the predicaments and dilemmas that emerge as an agro-pastoralist economy is being displaced and as related wisdom, knowledge, and community arrangements are eroding.

CHAPTER FOUR

"Father White Glacier"

Incommensurable Temporalities and Eroding Filial Bonds

The land is what gives us life. How can we abandon it?

—TSERING TASHI, TINGMOSGANG VILLAGE

Our elders would say that whatever bad happens, the land will always look after us. The land is perhaps difficult to take care of, but it gives you whatever you need. It will take care of your hunger, and it will look after your body.

—TSERING DOLMA, NYE VILLAGE

THE VIEW FROM TSERING ANGCHUK'S SMALL SHOP, WHERE NAMGYAL and I are having chai, is superb. As if watching a play from the balcony, we have choice seats from which to view the scene of farmers plowing the terraced fields down below. Higher up, looking over them, are the magnificent snow-covered summits bounding the valley. It is May and work in the field is in full swing in Ang, the upper part of Tingmosgang village. Because the village's fields are located at an altitude of about 3,600 meters, they yield only one harvest per year. Trapped in the inexorable cycle of the seasons, Ladakhis have to maximize the benefits of their arid land during the brief few months of warm weather before winter sets in again. During the warmest months of the year, there is not a day to spare. As the old Ladakhi saying goes, "Summer is the slave of winter."

I cannot help but feel that something is peculiar about the scene, though: while in some fields *dzo* pull implements of the kind that have turned this northern soil for generations, in others tractors rumble as they pull metal plows. The juxtaposition makes the former look like an anachronism. The plowman's exhausting work seems a bit futile beside the tractor's mechanized efficiency. As if reading my mind, Tsering Angchuk confides a different discomfort with the scene: "Isn't it a shame, using tractors instead of *dzo*." His remark encompasses so much of the turmoil that exudes from the villages of Ladakh today. Significantly, replacing *dzo* with tractors often means rejecting the constellation of religious and community arrangements embedded in Ladakhi agrarian activities.

Today, the trend away from a cooperative economy has altered the traditional cadence of farming. As people who own tractors rent out their plowing services from one village to another, it is often their availability that dictates the timing of agricultural activities, rather than the traditional, astrologically based agrarian calendar. Yet equally often, it is the lack of workforce in the villages that compels people to rent tractors. Tsering Angchuk's situation exemplifies this: with a measure of visible guilt, the retired serviceman confides that soon he, too, will have no choice but to rent a tractor, as his household lacks working hands, even though he has seven sons and one daughter: she has married into another household, two of his sons are serving in the army, two others are studying in different Indian cities, one is a monk, and another is a taxi driver in Leh. This leaves Tsering Angchuk and one of his sons with all the work. He had managed to finish plowing his fields the previous day, but only because he could afford to hire Nepali workers. His plight is shared by a great many farmers in Ladakh today.

Tsering Angchuk's discomfort with the mechanization of agriculture needs to be considered in light of the effect the displacement of the agrarian economy has had on social time in Ladakh. Drawing on Emile Durkheim's insights into the distinction between "the personal experiences in time from social time, that is, a person's sense of temporal orientation from his or her awareness of time as a category" (Greenhouse 1996:27), anthropologists have argued that time is a culturally flexible and socially manipulable resource (Gell 1992; Munn 1992). At the macro level, these manipulations have become a salient feature of modernity, bringing about a "time-space compression" (D. Harvey 1989:240) and changing the horizon of time, thus generating uncertainty (Comaroff and Comaroff 2001).

To capture the complexity of the experience of modern time, we need to focus on the encounters between different regimes of time and temporal representations. Institutions have become central actors in the contemporary mediation of time. For instance, the public rhetoric of macroeconomics and state planning contributes to changing the horizon of time, with a discourse that laments the past, describes the present as fluid, and renders the future as uncertain (Abram 2014; Knight and Stewart 2016; N. Mathur 2014). Although largely focused on economic, political, and bureaucratic social time, these anthropological perspectives on temporality generate insights into the coexistence of divergent regimes of time, its affective dimensions, and its flexible nature, which shed light on the changing agrarian times-cape of Ladakh. Both climate change and the new economic reality have contributed to changing the horizons of time: for many farmers, the lack of an agrarian workforce and repeated problems of water supply trouble the present and lead to a bleak prognosis for the future.

Yet as much as time is manipulated and shaped by institutions, people attempt to take control over time. Through the act of labor, temporal rhythms, representations, and technologies are mediated. But this mediation has become increasingly difficult to achieve under contemporary global capitalism (Bear 2015). Ladakhi farmers, too, have to negotiate with and mediate different temporal regimes. Capitalism increasingly dictates the rhythm of farm work, not so much because subsistence farming has been replaced by cash crops, but because the significance of farming activities has been superseded by other economic activities that take place outside the villages. Consequently, consideration for a sacred geography, which used to dictate the rhythm of farming activities, has become challenging for many families. For those who rely on the support of family members who live outside the villages, farming has to be accomplished as quickly as possible, often preventing the observation of the traditional rituals that punctuate the various steps of the farming season.

Besides the lack of workforce, farming in Ladakh today is also complicated by the recession of glaciers, which, according to Shammas, increasingly affects the flow of water in the streams. This problem is particularly acute in spring, when the fields have to be sown. Moreover, while farming relies on glacial meltwater, spring used to bring light rainfall that would help with the sowing. Yet rain is said to have become unpredictable, being at times in deficit, at times in excess.[1] During the spring of 2013, when villagers of Tingmosgang were confronted with a water shortage, they sought to address the problem by reviving a ritual meant to propitiate the protector of the main glacier of the village.

CYCLES OF INTERDEPENDENCE

In Ladakh, the farming season runs roughly from April to September. In this relatively short window of time, depending on the altitude, villagers harvest one or two crops, consisting mostly of a few varieties of vegetables

4.1   A *lha tho*, an altar dedicated to a local mountain deity

(potatoes feature prominently) and grains that grow well in arid soil, namely barley, wheat, and buckwheat. Each yearly cycle sees villagers trying to maximize the yield of the land before frost and low temperatures prevent all agricultural activity. Farming in Ladakh is not just about physically working the land, but also depends on people's relationship with divine beings. In traditional Ladakhi cosmology, agriculture requires both the physical work of lay people and the prayers and rituals of the monastic community. The land is the domain of the *sadak* (*sa bdag*), the lord of the soil, who has the final word on its use. The *sadak* are said to be quite capricious, punishing harshly those who fail to pay their due in the form of regular offerings (Dollfus 1996:31–32). The prosperity of the community therefore requires more than cohesiveness and hard work; the cooperation of the deities is crucial as well.

According to Tibetan Buddhist cosmology, spirits and deities are believed to inhabit various places, in particular the mountains and water sources within the vicinity of villages (Huber and Pedersen 1997; Salick and Byg 2012:451). These "gods and spirits of the landscape" are the "gods of this world" who guard the connection between moral values and the environment (Samuel 1993:161–66). The *sadak*, together with the *yul lha*,[2] the *lu* (*klu*),[3] and the *zhidak* (*gzhi bdag*),[4] are all territorial deities that feature in both Bon and Buddhist traditions (figure 4.1).[5] These deities are associated with geographical features and are critical to how landscapes are perceived. They form part of a complex of ideas about divination, spirit-mediums, and good and bad fortune, thus playing a central role in the pragmatic matters of their world. While the local gods are generally benevolent, they are also easily offended. Keeping on good terms with them and rectifying relationships when needed is the domain of ritual specialists. Rituals that aim at placating these deities have long been a central preoccupation of villagers in Tibetan Buddhist territories. And for good reason: it is believed that an unbalanced relationship with the deities can bring about diseases or natural catastrophes of various kinds (Kapstein 2006:209–11; Samuel 1993:179–92; Snellgrove 1967:12–13).

Fundamental to achieving prosperity, the acknowledgment for the interconnection between local spirits, human beings, and the environment is periodically enacted in ritual practices that bring together members of the community and specialists, such as monks, who engage the realm of the deities of the land.[6] Each spring in Ladakh, one such ritual, *sa kha phye*, marks the onset of the farming season and authorizes the villagers to start working the land and perform other farm-related activities (Dollfus 1996). Literally meaning "the opening of the mouth of the earth," the ritual consists of symbolically opening the land with the plow. In each village, the date and time of the ritual are set by the *onpo*, or local astrologer, who through various calculations establishes the details of the ritual.[7] During the inauguration ceremony of the first plowing, deities are entreated by prayers and by the burning of juniper twigs. Protector deities are solicited to guard the earth and seeds, while forgiveness is asked of the earth spirits, it being their ground that will be plowed. In assigning specific roles to the members of the community, and in requiring the participation of every member, the ritual also serves to reiterate the importance of community cohesion. *Sa kha phye* is also, in many villages, the moment to pay a tribute to the *dzo*, the animal whose power is harnessed to plow the field. In order to demonstrate their appreciation, villagers beautify their *dzo* by oiling their horns with butter and brushing their coat; they also offer them a copious meal.

At the basis of the ritual, like many others in Ladakh, is a worldview that integrates society and nature. Such has long been the rhythm of agrarian life in Ladakh: a succession of tasks that engage people with the environment, intertwined throughout with community arrangements and permeated with religious beliefs. Each of these "cycles of interdependence" (Bastien 1983) is dictated by the immutable changes of the seasons. In the past years, in many villages, *sa kha phye* has however become less elaborate, being

4.2   An *onpo* (astrologer) consulting his map

increasingly relegated to folklore, something that is also true of many other rituals (Dollfus 2008:136). But if the rituals are increasingly losing their significance, many villagers in Tingmosgang are clear: the date set by the *onpo* to start farm work has to be followed, and it is a shame that people are increasingly failing to respect this today.

We end up staying late at Tsering Angchuk's shop before heading back to the Chonpa house, about 2 kilometers away, where we are staying. Tsering Angchuk insisted on driving us back in his ramshackle truck, but I politely refused, somewhat concerned about his ability to drive: as other visitors had joined our gathering at his shop, the mood became rather festive and chai was soon replaced by beer. But I soon regret turning down Tsering Angchuk's offer. In my inebriated condition, walking the downhill path to get back home feels like running a half-marathon. Judging by the unsteadiness of his gait, Namgyal is not doing much better. There are no habitations along the way, and in the evening dark the sounds of the night are unsettling. In my imagination, every crack and snap emerging from the surrounding darkness is a snow leopard or a wolf, ready to pounce.

The next morning, after a brief night's sleep, I awake to the creaking of the door, as Padma, the daughter of our host family, brings tea to our room. The tea is a welcome balm to chase the taste of beer still in my mouth. I add a spoonful of instant coffee to the cup. The result, a brownish, excessively sweet mixture, perhaps objectionable from a gastronomic perspective, provides me with the defibrillating effect I badly need on this cold morning—for even in May, Ladakhi mornings are quite cold.

Namgyal is in deep slumber at the other end of the room. The sun is barely up from behind the mountains, but through the window I can already see farmers working in the distant fields. The Chonpa house is located in a very narrow section of the valley, and the sun appears only in the late morning, then vanishes in the early afternoon. I notice a large grasshopper walking on Namgyal's arm. This is an unmistakable sign of spring: insects, which had disappeared with the cold, are now returning. When the grasshopper suddenly crawls onto his hand, Namgyal awakens with a start. He looks at the creature for a few seconds, then brings it over to the window, which he opens, before gently letting the insect out. Here, squashing an insect is rarely an instinctive reaction.

In the kitchen downstairs, Padma is cooking chapattis. The sweet smell of grilling wheat fills the room. Padma's younger brother and sister are getting ready for school. Seven-year-old Skalzang is almost done arranging her hair into the school-mandated two large braids, while her brother, Norbu, who is five years old, clumsily sets the collar of his shirt. In their school uniform—pants, shirt, crested sweater, and tie—they look quite formal. Carrying their enormous school bags, which are probably half their size, the two then run to catch the bus by the roadside. Each day, Skalzang and Norbu come home with enough homework to keep them occupied until supper, which in Ladakh is rarely before eight o'clock. The amount of time Ladakhi children spend with their schoolbooks is indeed impressive and often leaves them with few free moments to play outside.

Padma is busy packing lunch and making tea for the relatives who have come to help her mother, Lhadol, prepare the fields. Because he works in the army, Lhadol's husband spends only a few weeks at home each year. On the wall of the kitchen, a framed picture of him in uniform watches over the family. Lhadol, a strong, spirited mother of four, is not only raising all their children, but she is also handling all of the farm work, including rearing the sheep, goats, and *dzo*. Her life conditions are far from unique in Ladakh. Hers is the burden of so many women married to men with military careers. She is overloaded with work. Indeed, the economic restructuring of Ladakh has had a significant impact on farmers, women in particular.[8]

Luckily for Lhadol, her eldest daughter, eighteen-year-old Padma, can now share some of the workload. Padma had come back to the village a year earlier, after spending several years at a residential school in Punjab where she completed her higher education. She has been far from the universe of Ladakhi villages for a long time, and her disconnection from rural life is obvious. As is the case with so many young Ladakhis, Padma's education meant her being uprooted from the culture and customs of her community. Spending years outside their villages, returning only for occasional visits, many young Ladakhis today find themselves struggling with feelings of alienation. But Padma knows there is a way out. Every morning, well before the first glow of the rising sun, she studies in preparation for her entrance exams in pharmacology. If successful, she could land a job that would bring her to the city. The intensity of her efforts is a telling indication of the strength of her desire to escape village life.

◆ ◆ ◆

In the field in front of Lhadol's house, her relatives, seven women and four men, along with two hired Nepali laborers, are busy preparing the soil for sowing. Contrarily to several villagers, the group only started plowing a few days ago, having followed the date prescribed by the *onpo*. The previous day, Lhadol had worked on her husband's brother's land. Today, the family group moved to her fields. Farming in Ladakh is a family affair that offers opportunities to celebrate, to gossip, to tease, and to reunite. It is as much about celebrating family as it is about getting work done in the field. Although hard work is clearly being accomplished, the scene has a festive feel. Padma carries

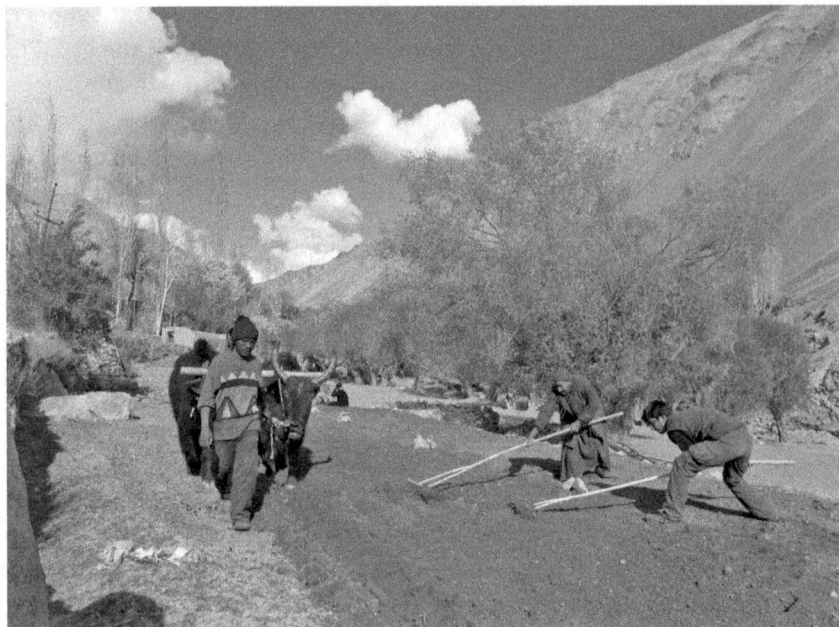

4.3   Plowing and sowing in Tingmosgang

out her part of the farming duties in the kitchen. Her job is to prepare food and make successive trips to the field to bring chai and butter tea to the group. She is given tasks other than working the land, for, as Ladakhi elders say nowadays, younger generations are "not strong enough" to do this type of work, as they spend so many years behind school desks.

Lhadol's family has not yet resorted to the use of machinery. A team of two *dzo* plow the field. The men take turns pulling them across the field by holding a ring connected to their noses. At the rear, a plowman strains to maneuver the harrow while expertly directing the *dzo* with specific vocal commands. Plowing is also hard work for the *dzo*; throughout the day, they are thanked for their effort with affectionate names.[9] But the task is also difficult for the plowman—which Namgyal ends up learning at his expense. To help the three plowmen, who are visibly exhausted, Namgyal tries to maneuver the *dzo*, but having little experience in farming matters, his attempts end up being a comedy performance for the benefit of Lhadol's family. This is one of those moments where, in his attempt to connect with village life and rural work, Namgyal is a remarkable "epistemic partner" (Marcus 2009:30). His curiosity leads him to ask questions about minute details that enrich my notebook. His quest to experience various agrarian activities also enables me to better grasp the challenges of returning to a rural environment, as have so many Ladakhis who have spent years away from home.

Once a section of the field is completed, the women level the soil with wooden rakes, before the men sow the seeds (figure 4.3). Now the field needs water, sun, and mild temperatures to prosper. The scene could be in any village in Sham at this time of the year. From one field to the other, it is like the performance of a musical fugue in which the same melody overlaps itself continuously, as people in one field begin to sing, followed by others in another field, and so on. It is a beautifully sonorous expression of cooperative farm work and a striking counterpoint to the sputtering of tractor engines.

◆ ◆ ◆

In the evening, we all gather at Chonpa *khangba*,[10] the home of Lhadol's sister-in-law, Dolma. Dolma's husband, too, works outside the village, but fortunately her two sons were able to free themselves from their jobs to come work the land. *Chang*, the local beer made from fermented barley,[11] is flowing at a steady pace. Spoonfuls of ground and roasted barley grain are added to our glasses, an additive that is said to intensify the effect of the alcohol. As we are offered glasses of *chang*, we each enact the *dzangs* (customary refusal), as required. In Ladakh, accepting an offer too quickly is considered rude. It is polite to first refuse an offer, once or twice, before accepting. Still feeling the aftereffect of the *chang* imbibed the previous night, I am not overly keen on indulging in drink. However, to refuse *chang* altogether would come across as a lack of solidarity with the group. The onset of the farming season is, after all, a festive moment, although my interlocutors repeatedly stress that such festivities are modest nowadays.

While everyone is chatting and laughing, I am captivated by the beauty of Dolma's kitchen. Ladakhi kitchens are somewhat akin to museums. The kitchen is where families demonstrate their wealth, by conspicuously displaying brass and copper cooking pots, bowls, plates, cups, and ladles, arranged on carved wooden shelves. These items are passed on from one generation to the next: as a general rule, as she leaves her natal household when she marries, a woman will receive the vessel her mother acquired at marriage; a man acquires the rest of the vessels by staying in the main family house with his wife. Ladakhi kitchens are not just places to prepare food and to celebrate; they are also repositories of genealogical biographies.

Voices at the door bring me back to the present. Dolma's sons are going back to Leh, where one is a taxi driver and the other an engineer employed by the government. Their time in the village, spent attending to agrarian work, meant days of lost business for one and requests to supervisors for leave of absence for the other. This is what farming means for many families in Ladakh today: trying to maintain a balance between work obligations and family responsibilities. More broadly, it is also about balancing individual aspirations and moral duties. A sudden melancholy comes over Dolma's face as she watches her sons go. Plowing is over and, with it, the reminders of life as it was in the old days, when subsistence united family members, rather than keeping them at a distance.

The difficulties of agrarian life experienced by the Chonpa family are shared by many agrarian families in Ladakh today, affecting older and younger generations alike. For some families, farming is the main source of subsistence, and replacing what has normally been unpaid cooperative work among family members by paid labor is becoming a pressing issue. Others, like Lhadol and Dolma, must adjust to off-farm employment. But regardless of the major economic changes the region has undergone, family farms remain the primary producers of food (Dame and Nüsser 2011:186). Moreover, despite the availability of imported and subsidized foods, winter in Ladakh without access to farm produce does not make for a nourishing diet and can become extremely costly.[12] This is why a great amount of time is dedicated to preparing for winter.

Farming the land also remains a moral obligation. Many villagers in Sham are discontented at seeing that today some people are "abandoning their land." The idea that one should never abandon the land but should continue to care for it despite the vagaries of agrarian work is a powerful one for Ladakhis, something, along with taking care of domesticated animals, not entirely remote from Buddhist virtues. The way Ladakhis speak about this care for the land is often framed as a moral responsibility: the land is depicted as a mother who cares for the subsistence of its children, whose duty, in turn is to reciprocate by providing care through farming. The community stigma associated with the lack of interest for one's land is such that some villagers believe that today "people are farming just not to be looked down upon by others." But the reality is that for many farmers today, especially for those who lack the support of their immediate family, farming has become problematic; for this reason, in Tingmosgang, many have stopped cultivating the lands that are in the upper part of the village, which are more labor-intensive to work and less productive.[13]

In the face of the shrinking agrarian workforce, some are resorting to renting tractors to plow their fields. These tractors are generally owned by entrepreneurial Ladakhis who, pensioned from the army or the civil service, have decided to invest in machinery they can rent out. Along with the increasing reliance on a labor force that lives outside the village, this contributes to changing the agrarian timescape. As people who own tractors rent their plowing services from one village to another, it is increasingly their availability, rather than the date set by the *onpo*, that dictates the timing of agricultural activities. As a result, many villagers no longer observe the traditional, astrologically based agrarian calendar, to others' great dissatisfaction. This is often expressed as a lack of community solidarity and a consequence of "people being in a hurry for everything." As one elderly Ladakhi woman told me, "Today, people are farming just for the sake of farming," implying the erosion of the social values and religious arrangements once attached to agrarian work.

THE OBSTINATE PROTECTOR OF THE GLACIER

The following morning, when I pick up my toothbrush by the window outside, near the spot of our morning ritual of washing our hands and face and brushing our teeth, I notice that the bristles are frozen. The temperature had dropped below zero during the night, freezing all traces of moisture, including the puddles that line the path in front of the house. Ominous clouds cover the sky. Here and there, timid rays of sun barely pierce the heavy blanket. The day does not seem promising. I draw Namgyal's attention to the inclement weather: "Do you think the family's crops are going to be fine?" I ask. Sitting on a rock near the channel in which flows a small stream of water, Namgyal seems quite annoyed by the cold. Usually fairly chatty, now he just warms his hands by pressing them to his mouth and blowing air. "It surely does not seem like a good sign," he remarks tersely. Obviously, Namgyal is longing to go back to Leh where he can sit by the *bokhari*.

Talking with farmers that day confirms our impressions. The meteorological conditions are relatively adverse at this crucial time of the farming season. Yet concerns about the cold temperature and cloud cover center not so much on crops potentially freezing—barley resists frost surprisingly well, farmers explain; rather, these conditions signal a delay of the seasonal melting of the glacier. Without a marked improvement in the weather, the seeds will lack water, thus jeopardizing the whole farming season.[14]

Tingmosgang is laid out in a distinctive Y shape that follows the branches of the local stream. On the west side, leading toward Tia village, the stream connects to two large glaciers, Onpo Kangri and Shali Kangri. To the east, it is fed by two glaciers of considerably smaller size, Kangri Nyingpa (old glacier) and Kangri Soma (new glacier). Water scarcity has always been an issue on the eastern side. Perhaps for this reason, Tingmosgang villagers have traditionally performed a number of practices that aim to monitor and nurture these glaciers. Although their efficacy can certainly be contested on scientific grounds, such practices testify to a close connection between the villagers and their glaciers, one that, as I suggest below, is eroding. In the recent past, the preservation of glaciers was a community responsibility (see the introduction). Human responses to the difficulties posed by the water supply in Tingmosgang have also long involved interactions with other-worldly powers. This is what I learn that day from Nawang Gyaltson, a man in his seventies known by his fellow villagers to be well informed in religious matters.

We meet Nawang Gyaltson on our walk through the village. He is sitting with another man, Sonam Puntsok, on the hillside next to the road eating dried apricots and watching over a flock of sheep that are grazing on a small patch of grassy land. The sheep belong to Nawang Gyaltson, who has much to say about his fellow villagers' eroding respect for the land and lack of consideration for the local deities. As we start asking about the current cold weather, Nawang Gyaltson is categorical in his diagnosis: somewhat warily, as if as a foreigner I embody the discourse of science, he cautions me, "It is not because of what some call 'climate change' that we have problems with water, but because people have changed and they no longer care for the *lha* and the *lu*." His comment strikes me, for it is the first time I have heard someone from his generation refer to "climate change." According to Nawang Gyaltson, the water problems facing the villagers are not unusual, for, being located in the shade of the mountains, the glaciers of Tingmosgang are prone to melt slowly in spring. But this geography alone does not explain the endur-ing problems with the water supply in Tingmosgang. Nawang Gyaltson—like his fellow villagers, as I later learn—believes that the *zhidak* of Tingmos-gang, or the guardian of its main glaciers, is a stubborn spirit, who refuses to let villagers carry out their farming activities unless they pay the necessary tribute.[15]

The belief that water problems are the result of a *zhidak*'s stubbornness assumes an interconnection between the realms of humans and the gods, as well as between the sacred, environmental, and human realms. In this world-view, the *zhidak* does not hand out favors: instead, the spirit obstinately requires that villagers acknowledge that interdependence through rituals.[16] In the past, according to Nawang Gyaltson, as a preventive measure, a ritual known as *skyin jug* was systematically performed in order to appease the protector. But over the past two decades this cautionary practice has been gradually put aside. In recent years, as the water supply has become increas-ingly problematic, some villagers—among them elders and "those who don't benefit from salaries," as Nawang Gyaltson puts it, suggesting a class divide among farming families—are seeking to revive the ritual. Most of these people are of the older generation because, in the words of Nawang Gyaltson, "young people have no interest in this type of activity." It is the same lack of interest, in his view, that has apparently led to the abandonment of the ritual over the years. "See this steep mountain," says Nawang Gyaltson, pointing at the landform that bounds the village. "How can old folks like us climb this?" he asks. *Skyin jug* has to be performed on the top of this summit, from which it is possible to see the glacier. From the place where we are sitting, the mountain looks like a wall rising out of the landscape. It is no doubt a challenge to ascend. That spring, according to the *lotho* (*le tho*), the Tibetan almanacs used to foretell weather, no rainfall was predicted to come anytime soon; as a result, this potential fix, the performance of *skyin jug*, was now being seriously considered.[17] Even a sparse rain would at least dampen the soil until there was a more generous flow of water in the stream.

"But why don't people climb this mountain instead?" asks Namgyal, pointing at a mountain on the opposite side of the valley that, from our standpoint, seems also to provide a view of the glacier, and with a less oner- ous climb. "That is the cemetery of the Balti," explains Nawang Gyaltson.[18] "*Azhang-le*, you mean, as in dating from the war?" I ask. "Yes, during *arthalis*, many Balti were killed here. If you go, you will see many graves." He pauses for a moment, then continues: "Many Balti died at that time. It was horrible. I was young, I still remember." He pauses again. "It was a very horrible time." We all remain silent, observing the mountains. "In any case, you cannot see the glacier at all from there. It is well hidden!" he exclaims, as if to avoid the grave subject of the war.

As though it had been written into the plot, that same evening brings another source of insight. One of Lhadol's neighbors, Palmo, who regularly drops by to visit, has invited Namgyal and me for dinner. A retired school- teacher in her sixties, Palmo is a beautiful woman, with long grey hair and luminous green eyes. She has a contagious smile and is always making jokes, but she also complains about how boring village life is. She is still sour about having come all the way from Hemis to marry a man she fell in love with four decades ago, a man who, she once explained, spends most of his time in Leh with his son now that he is retired from the army. Her loneliness probably explains why she spends hours every day at Lhadol's house. Only when Namgyal and I arrive at her house for dinner do I realize that Nawang Gyaltson's companion earlier that day, Sonam Puntsok, is her husband. He had been largely silent during our prior conversation, and that evening Sonam Puntsok apologetically confides, "I was afraid you would ask me questions about the culture of Ladakh earlier on today. I was in the army; I spent all my life outside. I know nothing about that."

In the days that follow, the weather remains inclement. The feeling of spring we had briefly enjoyed in the previous weeks seems long ago. This does not compel farmers to stop their work in the fields, but in conversations many express concern about the weather conditions. When I ask around, the villagers are still unsure if *skyin jug* is going to be performed. Indeed, gathering information on the ritual proves difficult. The ritual has not been performed for a few years, and villagers' grasp of the practice seems hazy. But most tellingly, no one seems to know *who* will perform the ritual. While many allude to the fact that only elders have knowledge about such matters, apparently no one wants to take the lead in doing something that they feel is the responsibility of the younger folks. The revival of *skyin jug* seems to remain only a proposition. I conclude that the best way to make sense of this discord is to meet the appointed caretaker of the local monastery, who would normally be involved in the ritual.[19]

• ◆ •

Perched on top of a rocky hill that splits the village in two, Tingmosgang's monastery is simply stunning in its setting. After a brief trek along the steep path that leads to the monastery, we arrive panting at the cell of the monk in charge, a man in his late thirties. He receives us in his room, a dark chamber that reflects the prescribed frugal lifestyle of his profession, with very modest furniture punctuated by scriptures and a small stove. He invites us to sit on one of the three beds covered with modest carpets that serve as seats for visitors. A young girl soon arrives with tea, biscuits, and dried apricots. The monk excuses himself in advance for not being able to spend much time with us: he soon has to return to the temple, where prayers are going on.

When I inquire about *skyin jug*, he informs us, to my surprise, that he has never heard of this ritual. Monks in Ladakh are posted on a rotational basis, he explains, and he was only recently transferred from Igu, where such practices do not exist. Even more puzzling, no villager has come to him in the past few days asking about it. In any case, the monk admits, "I am not well versed in rituals of this kind"; older monks would know better, he says, as the training of monks of his generation is oriented more toward textual Buddhism—knowledge of prayers rather than the performance of rituals. "Besides," he adds, "rituals of this kind are less popular nowadays."

On these words, the monk remains pensive, sweeping his fingers across the wooden table in front of him. His hands are soft and smooth, devoid of the many scars seen on the hands of so many of the farmers around. Undeniably, he has spent most of his life away from agrarian work. He then continues: "One of the major reasons [for the decline in these ritual practices] is also that these spirits do exist, but not in the solid form of flesh and blood like human beings. They are formless and invisible to the naked eye, so the common people doubt their real existence in our environment. Because of this and the pursuit of material gains and development, the modern people distrust and disbelieve the age-old religious traditions and practices and they neglect many aspects of religion. This leads to imbalance in our environment and society and thus causes indignation to all these *lha, lu*, and *zhidak*."

The monk's explanation suggests that the abandonment of *skyin jug* could be linked to the growing influence of the rational outlook on the environment. This trend also has other, more locally institutionalized ramifications. In Ladakh, as in other parts of the Himalayas, high lamas have for many years been pushing an agenda of orthodox Buddhism and often suggest that some rituals deemed heretical be abandoned.[20] But this equally points to the generation gap regarding religious knowledge, with ritual experts now clearly seen as belonging to another generation, if not downright as a vanishing resource. This is something Ladakhis are well aware of, for as I later learned, the reason nobody went to see the caretaker of the monastery to ask about *skyin jug* was that they knew it would be in vain, since he is of a younger generation of trained monks who know little about traditional rituals. In short, the ethical responsibility to perform *skyin jug* as a means of recognizing the role of the *zhidak* is difficult to fulfill with Buddhism becoming increasingly orthodox.

We leave the monastery feeling that we are trying to solve a puzzle that is becoming more complex by the day. Each piece we have gathered provides, it seems, only a partial view of the ritual that villagers were attempting to revive. That evening when we return to the Chonpa house, Lhadol bursts into laughter upon seeing our disappointment. "You are breaking your heads so much with this thing!" she exclaims. Seeing how busy she is from morning to night, I wonder for a minute about what I am trying to accomplish. "Why don't you go see *meme* Nyima?" she suggests. *Meme* Nyima had been the appointed monk of Tingmosgang monastery for many years, and Lhadol had recently heard that he was now living in the monastery of Likir, about forty kilometers away. "He probably has more to say about *skyin jug* than anybody here," she says.

The massive, centuries-old monastery of Likir, which belongs to the Gelugpa sect of Tibetan Buddhism, dominates the village below, which is spread out inside a long, wide valley (figure 4.4). Adjacent to it stands an equally—if not more—imposing structure, a twenty-three-meter-high statue of Maitreya, the future Buddha of this world in Buddhist eschatology. Its construction, completed in the late 1990s, reflects the commodification of religion that has taken place over the recent decades in Ladakh, as past practices of devotion are seemingly being replaced by conspicuous constructions. Flamboyant with vivid colors and visible from a fair distance, the statue grates on the sensibilities of some, as do the many oversized prayer wheels built with increasing frequency in the villages. Many Ladakhi elders see a paradox in these constructions: in their view, today, people are eager to show that they have money to build religious monuments, but "they pray with empty hearts," or for show, without true humility.

In one of the monastery's two assembly halls, whose walls are adorned with colorful paintings of religious deities, we meet a young monk in his late teens. His blue-tinted sunglasses and basketball shoes clash with the prescribed asceticism of the monastic life. Like so many monks of his age, he is probably feeling the pressure of having to make the difficult choice between continuing the religious vocation or turning toward secular life. Upon our request to see *meme* Nyima, he smiles nervously, then shows us the way to an old section of the monastery that houses elderly monks. With their antique appearance, the monks' living quarters are quite in harmony with their occupants: both are well advanced in age. The staircase is in disrepair and the walls are crumbling. On the terrace, a monk sits on the floor painting a wooden door, while another sands a wooden table. Although falling apart, the elderly monks' quarters seem to be the liveliest section of the monastery. Our guide is in fact an increasingly rare example of youth in a vocation that is undergoing a steady decline. The young monk confides to us his apprehension, for *meme* Nyima is apparently quite a persona. He leaves us in the inner courtyard while he goes knocking on *meme* Nyima's door. After a few minutes, the young monk comes back and whispers, "*Meme* Nyima is coming." He then retreats to the hallway, where, succumbing to his curiosity, he stops to listen to the exchange that is about to unfold.

Meme Nyima arrives on the terrace and complains half-heartedly, with a husky voice, "You'd better have a good reason to disturb me during my nap." He is in his eighties and wears a threadbare maroon robe; gray stubble covers his face. "I know nothing, and I forgot everything. What do you want

4.4   Monastery of Likir

me to tell you?" he asks. By now, I am already familiar with this type of reception, a typical approach of monks of his generation. There is an assumption that, as a foreigner, I am seeking information on orthodox Buddhism and scriptural knowledge.[21] When monks of *meme* Nyima's generation received religious training, philosophical studies of Buddhism were uncommon. Monks had to study in Tibet—a situation that changed after the 1959 Tibetan Uprising—and their religious training was typically oriented only toward ritual practices, rather than scripture. As a *geshe*[22] of *meme* Nyima's generation once told me, many elderly monks in Ladakh today can "perform the rituals, beat the drums, and hit the cymbals," but "they don't know the meaning of the texts." The opposite is equally true, however: today, aspiring monks are much more likely engage in the philosophical study of Buddhism, with ancient rituals now seen as deviating from Buddhist orthodoxy and no longer a central concern of religious curricula. This explains why *meme* Nyima does not know the texts—which is what he indirectly expresses when he says he "knows nothing"—and, conversely, why the caretaker we met at the monastery of Tingmosgang is not well versed in rituals that do not fall under orthodox Buddhism.

With this in mind, I inform *meme* Nyima of the reason for my visit and my interest in *skyin jug*. To my relief, his expression changes and he becomes enthusiastic, as if I have revived long-forgotten memories. In the past, he tells me, *skyin jug* was systematically performed in spring to ensure that there would be sufficient water for the crops. This invoked a complex ritual apparatus requiring the participation of monks, musicians,[23] children, and lay people, all of whom would climb the steep mountain at the upper part of Tingmosgang village. The ritual, which he describes as being intensely affectively charged, unfolded on the summit, which allows a full view of the main glacier of the village and which is the abode of the local *zhidak*.[24] At some point, entranced worshipers appealed to the *zhidak* by repeatedly shouting the following incantation:

> Father white glacier, ju hey![25]
> Mother, *mapam* lake, ju hey!
> *Zhidak* of the village, sacred owner of the land, ju hey!

Mapam refers to Lake Manasarovar in Tibet, at the foot of Mount Kailash, the holiest peak for Tibetan Buddhists. The lyrics, which remind us that Buddhist Ladakhis see glaciers as male and lakes as female, are a form of respectful salutation to the father, the mother, and the *zhidak*.

The *skyin jug* ritual appealing to the *zhidak* of Tingmosgang mimics a ritual of the same name performed during Ladakhi wedding ceremonies.[26] Expressing a bride's lament at leaving her family, *skyin jug* is a form of poetry normally performed exclusively by women (Aggarwal 2004:133).[27] It is performed during the elaborate leave-taking ceremony that marks a bride's departure from her home, as she leaves with the groom's party.[28] As her departure becomes imminent, the bride wails hysterically, accompanied by her friends. After a series of highly codified rituals that span a few hours, the bride performs the last part of the ceremony, during which she recites stock phrases and cries loudly. The verses of *skyin jug* are partly improvised by each performer. Overall, however, the chants reflect Ladakh's predominantly patrilocal postmarital residence patterns. As a general rule, a bride performing the ritual sings about her family members and the pain of leaving her natal home. Her relatives and friends accompany her in her sobs, making for a chorus of wails that fills a house with the sounds of lament. It is said that the daughter cries out of sadness over leaving her close ones and because she wants her family to remember her, while her kin cry to implore the family-in-law to take good care of her.

The *skyin jug* traditionally performed on the mountaintop in Tingmosgang echoes these tropes. The community takes on the role of the bride leaving her natal home and respectfully salutes her father and mother. In affirming their filiation with a glacier and a lake, the villagers are affirming their filiation with the broader natural world. Because they are sources of fresh water, life's most fundamental resource, the glacier and the lake symbolize a father and a mother who take care of their children. Through the ritual, villagers acknowledge that they live under the patronage of the local *zhidak*, without whom they would be at a loss.

In Tibetan Buddhist areas, many songs and rituals, including those related to marriage, include glacier imagery and sometimes integrate liturgical elements. In these songs, humans acknowledge their subordination to the elements of the physical landscape and the forces that animate them, reiterating the crucial role the ice peaks play in ensuring fertility and, thus, the continuity of the community. Although these songs do not always explicitly emphasize a filial connection between glaciers and people, they do portray glaciers as fulfilling a similar protective and life-giving role and as being the patrons of human communities.[29] Geoff Childs (2004:101), for instance, reports how mountain imagery is evoked throughout wedding ceremonies among Tibetans of Sama in Nepal. The following excerpt is part of a song performed by the groomsmen as a bride departs from her natal home:

> Ama layho!
> The snow peak up there is the eldest,
> The earthen mountain below is younger.
> Earthen mountain, please protect the snow mountain!
>
> Ama layho!
> The earthen mountain up there is older,
> The rocky cliff below it is younger.
> Earthen mountain, please protect the rocky cliff!
>
> Ama layho!
> The earthen mountain up there is older,
> The rocky grassy hillock below it is younger.
> Earthen mountain, please protect the grassy hillock!

In these verses, vertical layers of mountain topography are used "as an allegory to highlight cultural ideals about social relations with the family" (Childs 2004:102), with each layer—snow peak, earthen mountain, rocky cliff, and grassy hillock—symbolizing one of the four generations within an extended family. In these verses, Childs explains the snow peak—the highest and most venerated part of the mountain—represents the grandparents' generation. The second-oldest generation, the middle-aged children of the grandparents, are the nucleus of the multigenerational family and are represented in the song by the earthen mountain, that is, the solid core of the mountain. The allegory is multilayered: the middle-aged adults have the bodily strength to sustain the economy of the household, they have knowledge of farming and herding, they have sagacity, and they have acquired political knowledge throughout the years. For all these reasons, they are called on to provide protection to the other layers of the mountains: to provide support to aging parents (the snow-covered peaks); guidance to their children, who are coming of age (the rocky cliffs); and care for their grandchildren, who are making their entrance to this world (the grassy hillocks). Both the verses reported by Childs and the ones from *skyin jug* associate elements of the mountains with family relations, which constitutes a central dimension of the organization of life in the Himalayas.

According to *meme* Nyima, the *skyin jug* ritual would conclude with all the participants shouting, "Kiki So So Lhargyalo," the call having as much relevance in ritual as in warfare, as we saw in the previous chapter. While discussing *skyin jug*, *meme* Nyima extolls the virtues of the ritual, as did many farmers I talked with in Tingmosgang, insisting that water flows from the mountains within a few days and sometimes within just a few hours of the performance. Listening in on his story, another monk, who is sitting with us on the terrace, glances skeptically at *meme* Nyima. Obviously vexed by his fellow monk's disbelief, *meme* Nyima loudly challenges anyone who doubts his words to ask "the nun of Ruth house," as "she has seen it with her eyes."[30] Old age has not dulled *meme* Nyima's verve, and he is energetically indignant when his expertise is questioned. With great zeal, he then exclaims: "I am not exaggerating. I really did it back then, *konjok*! I did it myself, I swear, I truly brought water along with me from the glacier! I did it for the villagers, it was my responsibility!" Notwithstanding his insistence on the ritual's effectiveness, *meme* Nyima confides that *skyin jug* for the *zhidak* of Tingmosgang has not taken place in several years. The problem, he explains, is that to be effective, the ritual requires the mobilization of the whole community. "But if there are only old people interested, who is there to climb to the top of the mountain?" he asks, adding: "And me, I am old."

Although it is easy to doubt the efficacy of *skyin jug*, it is worth mentioning that weather-making specialists can be seen as ethnoclimatologists—that is, they have great skills at reading the sky, the stars, and anticipating the weather to come. Accordingly, they decide the right date on which to perform the ritual.[31] But for the family members who are only part-time dwellers of Tingmosgang and who must take a break from their current employment to help with farming, waiting for the right day to perform the ritual is not an option. The decreased stamina that comes with old age also poses an obstacle, for performance of the ritual involves climbing a steep mountain.

The case of Tingmosgang illustrates the importance of community mobilization for the ritual to be performed. When communal bonds are broken and farming shifts from a cooperative to an individual approach, such solidarity proves elusive. In the absence of community unity, the ritual simply cannot be performed. Besides, it is today increasingly difficult to find ritual specialists. *Meme* Nyima was once the central figure in a practice that villagers believed was saving them from unfavorable harvests, thus ensuring their prosperity.

• • •

In the end, the *skyin jug* ritual was never performed. The villagers waited anxiously for days for the rain to come or for the weather to improve. This tension lasted until the end of May, and fortunately the crop was not compromised, although according to many, it was not a good year. Several factors may account for the failure to revive *skyin jug*. For one, with the rationalization of religious practices, and since *meme* Nyima has no successor, finding an expert to conduct *skyin jug* proved a problem. Moreover, like all Buddhist weather-making ceremonies (Mills, Huber, and Pedersen 1998), *skyin jug* used to be financed locally by people who wanted to protect their fields. Taking a functionalist perspective on the ritual, we could link its unsuccessful revival to the displacement of agriculture as a central economic activity.

But this sheds only partial light on the failure to revive *skyin jug*. Despite the access to monetary gain through off-farm employment, family farms remain crucial to the production of food. Moreover, if most elders are categorical about the virtues of *skyin jug*, younger Ladakhis mainly speak of it as belonging to elders' knowledge, without interrogating its pertinence. As Lhadol, our host in Tingmosgang, told me, "Whether it works or not, trying cannot hurt." Her attitude is typical of what I encountered among the villagers of her generation.

Other elements also account for the unsuccessful revival of *skyin jug*. A few days after my visit to *meme* Nyima, I again meet Nawang Gyaltson, one of the most knowledgeable persons on *skyin jug* in Tingmosgang. During our conversation, Nawang Gyaltson praises *meme* Nyima's talent for entreating the *zhidak*: "Some years the water would flow down into the stream the same evening, by the time the monk would come down from the mountain," he says. "But why," I ask, "is *skyin jug* not being performed this year, despite the water problems?" Nawang Gyaltson answers by evoking disparate reasons, which together suggest that overall, organizing the ritual has not been the object of concerted effort among villagers. "Some people talk about it here and there," he says, "but it seems like it is becoming as much a headache to organize it as is the lack of water itself! In any case, where do you find someone like *meme* Nyima these days?" In a cynical tone, Nawang Gyaltson laments, "Young folks always expect us to do everything, but they don't realize that doing *skyin jug* is also paying respect to the *zhidak*. With this attitude, it is not surprising that we have water problems!" This comment illustrates the perceived malaise of a generation in the face of traditional social responsibility: if organizing *skyin jug* requires effort, in Nawang Gyaltson's view, it should be taken care of by young, able bodies. Moreover, if for Nawang Gyaltson the performance of *skyin jug* is a form of moral recognition of the *zhidak*, others in the community, as Lhadol's comment suggests, do not recognize this moral imperative. Rather, *skyin jug* is seen as a fix, devoid of moral context.

It seems that the failure to mobilize for the ritual was also, to a large extent, the result of conflicting perspectives on agrarian time. Traditionally, agrarian time is informed by an acknowledgment of the interdependence between the laity and the deities. But among family members, farming is today increasingly superseded by other economic activities, which means that the labor-intensive steps of the farming season, such as sowing, must be accomplished quickly. As a result, the rhythm of capitalism and the rhythm of sacred geography prove to be incommensurable. Organizing the ritual, looking for an expert, and participating in the ritual itself are all aspects that would normally be the responsibility of younger people; yet today, this generation is struggling to support farming family members while meeting their own work obligations, often outside the villages.

CONCLUSION: ETHICS AND THE AGRARIAN LAND

The failure to revive the *skyin jug* ritual for the *zhidak* of Tingmosgang signals the erosion of the filial link that Ladakhis have long imagined unites them to glaciers. Ultimately, it also signals a changing ethics of the land. Farming in Ladakh has traditionally been marked by rituals that attend to the interdependence between the laity, deities, and nature. Respect for the ethics of the land, whether by following the agricultural calendar set by the *onpo* or by actively participating in farm work, is today complicated by the different temporalities that define agrarian work. Both climate change and the new economic reality have changed the horizons of time for many

Ladakhis, making agrarian work increasingly difficult. Farmers today lack confidence in the present, which has become ever more fluid and unpredictable, and in what life will bring in the future. People describe farm work and village life as no longer what they once were, the values, rituals, and practices once defined as crucial to the cycle of the seasons having seemingly eroded as people increasingly opt for off-farm employment.

The uncertainty that characterizes the present and the future brought an additional layer of intent behind the desire to revive *skyin jug*. While in the past the ritual was systematically performed as a preventive measure, now its intended function was to address a problem of water supply, likely attributable to climate change. If the motivation for the performance of the ritual was, as in the past, a means to restore a cosmological order, it was also an attempt to mediate time, by putting an end to a period of cold temperatures. In this context, *skyin jug* was a "technological fix," or an "ethically and affect-driven practice" (Bear 2014:86) that acquired a new dimension, that is, countering the effect of climate change through the manipulation of time.

Rituals have been described as "the ultimate human emplacement structure," which "more than any other human institution . . . reflects and shapes the order of things" (Thornton 2008:174). Accordingly, the performance of *skyin jug* for the *zhidak* of Tingmosgang reiterates a fundamental disposition of Ladakhi society, which in the Tibetan Buddhist cultural construction of order posits that humans live under the patronage of glaciers and their associated deities, which provide life-giving water. Perhaps more tellingly, *skyin jug* used to be based on "landscape kinship," namely the belief that the landscape of Ladakh is inhabited by a deity with whom villagers have a filial link. Prioritizing a cultural rather than biological approach, Marshall Sahlins (2013:20) defines kinship as "a manifold of inter-subjective participations" and a "network of mutualities of being." Though this definition does not specify that participants in the system share a common substance, it nonetheless suggests that kinship connects conscious beings. But the ritual of *skyin jug* posits a filial link with a different type of being, a genealogy that links glaciers and humans. Kinship as a system has its own temporal dimension, which rests in the sequentiality of generational reproduction (Feuchtwang 2013:284). That the revival of the ritual was unsuccessful signals the rupture of the filial bond between the laity and the deity and suggests the erosion of ideas about the hierarchy between glaciers and humans. Ultimately, the failure to perform *skyin jug* highlights a mode of dwelling in the world that is transforming.

CHAPTER FIVE

Searching for *Aba* Stanzin

On the Predicament of Herders

When a *dzo* gets old it's the slaughterhouse
When a man gets old it's the prison cell

—TIBETAN PROVERB

I was a shepherd as long as my body allowed.

—SONAM ANGDUS, SASPOL VILLAGE

There are not many goats and sheep these days because people have
become bad and are not interested.

—NAWANG STOBDAN, TINGMOSGANG VILLAGE

IT IS MIDSUMMER. WE ARE ON A BUS TRAVELING THROUGH SHAM. A group of old men is chatting and teasing one another. The sound of the *dramyin*, the traditional Ladakhi lute, and high-pitched voices compete with speakers blaring Bollywood-inspired Ladakhi music. The bus is overladen with exhausted people, gas cylinders, bags full of goods, and assorted pieces of luggage. Battered by years of travel on some of the world's most hazardous roads, the old machine struggles to climb yet another hill, making sounds that grow less reassuring by the minute. As we speed downhill after each peak, the harsh sound of the engine's exhaust tears through the silence of the mountains. Each curve makes me feel as though I am on a roller coaster. So far we have stopped three times, the driver and the conductor disappearing under the bus each time for a round of repairs. This is transit in Ladakh.

Mobility in the mountains relies on a system of privately owned buses, whose years of service have taken a toll on their reliability, resulting in perpetual delays for those who depend on them. A trip often ends up being prolonged due to the customary flat tire, a consequence of the poor road conditions, or mechanical difficulties. This time, there is apparently a problem with both the brakes and the transmission. Cliffs plunge hundreds of meters downward on one side of the road. Here and there, the lunar landscape is strewn with white army parachutes pitched as tents for workers who have come all the way from Bihar and Nepal to carve roads in these mountains.

Sculpted by water and the imperceptibly slow journey of glaciers over millennia, the Ladakhi landscape we traverse makes for stunning scenery. On our way to Saspotse, we pass an impressive line-up of trucks being loaded with military equipment, goods, and petrol before the roads are closed for winter. The driver takes a curve too quickly, and passengers are thrown against one another. The road then opens onto a scene worthy of a National Geographic documentary: a wild predator is feasting on the remains of a freshly killed cow by the roadside. When the man behind me screams "Kishang!" the passengers rush to the windows. A dog-wolf hybrid, the *kishang* population appears to have multiplied in recent decades, posing a regular threat to livestock.[1] Behind us, a worried passenger asks fellow villagers if they sent a watchman to guard their cows in the pasture that morning. "Yes," answers one of them in a reassuring tone. Then a sarcastic voice from the back of the bus shouts, "Yes, they must have sent a watchman over here too, and look what happened to their cows." Silence follows, everyone's thoughts turning to the safety of their livestock.

Thought to have exploded in large part because of easy access to food waste from military installations and road builders' camps, and thanks to wildlife protection laws, this predatory canine population is a blight for livestock herders,[2] who struggle to keep their animals through thick and thin. The threat *kishang* pose to farm animals is only one among a number of recently emerging predicaments faced by herders that are altering the pastoralist landscape of Ladakh.

Long defined as a static object of human observation, landscape has for the past two decades attracted growing interest among scholars in a variety of disciplines.[3] Landscape is a fundamental feature of people's existence, and through a study of the affective aspects of topography, this scholarship attempts to transcend the representational character of former approaches by moving away from notions of landscapes as "framing convention" (Hirsch 1995:1). Landscapes, indeed, are imbued with cultural significance, functioning as repositories of collective memory, mythology, and social identity

(Basso 1996; Thornton 2008). In addition to emphasizing local experiences, this scholarship has shown landscape practices to be intertwined with various forms of power (Fairhead and Leach 1996; Mitchell 2002; Moore 2005). Whether through political forces or environmental processes, landscapes are not fixed, and their changing nature has cultural resonance (Cruikshank 2005; Rasmussen 2015). In sum, "landscape blurs the boundaries of objective reality and subjective human experience" (Willow 2011:263).

But how are these landscapes produced? What networks of practices lead to their emergence? If landscapes are culturally made rather than emplaced, as anthropologist Laura Ogden (2011:26–27) points outs, an anthropocentric perspective on their constitution may obscure the relations with nonhumans that are sometimes central to their production (Ogden 2011:26–27). Ogden's reflection that "our relations with non-humans produce what it means to be human" (28) is particularly apposite to the tracing of a landscape ethnography of pastoralism in Ladakh. Scholars have come up with a range of metaphors— assemblage (Deleuze and Guattari 1991), cyborg (Haraway 1991), hybrid (Latour 1993)—to represent the intermingling of society and nature and to emphasize the inextricability of humans and elements of the material, semiotic, or natural environment from the relations between them. The pastoralist landscape, too, can be seen as an element of socio-nature, produced through the entanglement of domesticated animals, people, and mountains.

The productive dimension of landscape is not limited to the relationships through which they are constituted, and participation in the collective of relations that leads to a landscape's production also transforms humans. The idea that humans are produced through their interrelation with nonhumans provides an apt translation for how Ladakhi herders correlate forms of selfhood with herding work in the mountains. For Ladakhi herders, humans have a moral responsibility to care for animals, and the growing abdication of this responsibility is for them a symptom of a changing moral attitude.

HERDING AS A WAY OF KNOWING THE MOUNTAINS

In 1887, Francis Younghusband crossed the Mustagh Pass, which marks the watershed between India and Central Asia and forms part of a challenging route from Yarkand to Leh. At an altitude of about 5,800 meters, within sight of K2, the world's second-highest peak, Mustagh is a formidable challenge. Younghusband is reputed to be the first European to have accomplished the crossing, a feat repeated only twice since then (Rizvi 1999:26).[4] This was the time of the "Great Game," the strategic rivalry between the British and Russian empires vying for supremacy in Central Asia. One of Younghusband's guides was a man from Baltistan named Wali, whose loyalty and skill made a strong impression on the English explorer. Following the hazardous crossing, which elicited great fear in the ranks of the expedition crew, composed of Ladakhis, Chinese, and Baltis, Younghusband asked Wali if he would be willing to guide him once more over the Mustagh. In his memoirs, Younghusband recalls: "He replied that he did not want to go, but if he were really required he would undertake to guide me; the only condition he would make would be that I should not look at a map. He had heard that Englishmen were rather inclined to guide themselves and trust the map rather than the man with them. If I was going to do that, I might, but he would not go with me. On the other hand, if I would trust him, he would take me safely over" (Younghusband [1904] 2009:105–6).

This passage, which demonstrates that outsiders have long relied on local knowledge to navigate the Himalayas, also reflects two fundamentally different ways of reading landscapes. What distinguishes Wali's knowledge of the mountains from that of "Englishmen" reading maps is its embodied nature. Maps, described as tools of colonial supremacy used to dominate landscapes and populations, constitute a particular mode of engaging a place: a map freezes knowledge in time, approximating the components of the physical environment as lines traced on paper. But for a guide like Wali, knowledge of a place is a dynamic process that emerges from bodily engagement. By reading the topography, he can arrive at an empirically informed estimation of what might lie behind a summit, what route is best to follow, and what dangers may lie ahead. Mountainous terrains, moreover, are not fixed, and maps cannot account for the myriad reconfigurations brought about by avalanches, rockfall, and other processes of change. For all these reasons, a guide like Wali, for whom the mountains were a lived experience, was right to be suspicious of maps. In Wali's way of knowing, people and land, the perceiver and the perceived, were not separate entities that could be considered in isolation: a landscape was a lived space.

But mountain guides' seemingly innate ability to read the mountains comes only from years of practice and experience. In Ladakh, one way in which people acquire the skills and knowledge necessary to read the mountains is through pastoralist activities, which they often take part in from early childhood onward. Tim Ingold's research on hunters and gatherers (2000, 2007) is particularly useful for understanding the key role that animals play in knowledge acquisition processes and landscape practices. Ingold rejects the long-held anthropological perspective that culture precedes knowledge acquisition, and instead foregrounds movement and exploration. "Through the practical activities of hunting and gathering," he explains, "the environment—including the landscape with its fauna and flora—enters directly into the constitution of persons, not only as a source of nourishment but also as a source of knowledge" (Ingold 2000:57).

The contrast between Wali's and "the Englishmen's" approaches shows different ways of knowing the mountains, one that gives primacy to embodied experience, the other to objectified data. It also gives us insight into how knowledge about the mountains in Ladakh may develop through pastoralist activities and the role played by animals in the shaping of this knowledge. Multispecies anthropologists have in recent years invited us to rethink the relationship between humans and nonhumans by calling into question the human exceptionalism that has long characterized their discipline. Thinking of animals, microbes, plants, and others as acting agents opens our analytical perspective onto a world in which nonhumans are more than receptacles for human action, but can also act on humans (Kirksey and Helmreich 2010).

In Ladakh, knowledge about the mountains develops by traveling the pasture with domesticated animals. The course of these movements is dictated partly by the animals as they forage for food and partly by the herders as they attempt to control their herds' movements. The work of herders involves perceptive sensorial observations, as one must remain alert to the ever-present threat of predators. Herders devote a lot of time to scrutinizing the landscape and listening to the surrounding sounds. To be effective, a herder must also keep a sharp eye on the flock to guard against animals running away or otherwise getting separated. Ladakhi lore serves to reinforce the necessity of alertness in the mountains: stories hold that lazy herders who fall asleep in the mountains are assailed by cruel *smanmo*, or female ghosts.

A herder's journey in the mountains is not the same as a journey along a fixed route: rather, it unfolds according to the direction and pace of the animals and the location of mountain resources. Ingold (2007:75–81) distinguishes between "transport," defined as the near-automated bodily movement through space between determined points, and "wayfaring," defined as the sensorial engagement with the environment through the act of taking unplanned routes across space. What distinguishes transport and wayfaring, for Ingold, is not the recourse to a vehicle but rather the modes of engagement with the environment and the perceptions they imply. The motion of the wayfarer, in Ingold's view, hinges on the "sights, sounds and feelings" that pervade his route (2007:78). Like wayfarers, herders in Sham experience a range of sensorial observations and stimuli that unfold along more or less unplanned routes.

Ingold's proposition that, through practical engagement with the environment, a person is constituted cognitively by acquiring knowledge, and that such knowledge may be undergirded by sensorial engagement through mobility, sheds light on how herders' participation in the production of the pastoralist landscape leads them to develop an extensive body of knowledge about the mountains. Yet this perspective remains largely focused on the practical side of the equation, such that engagement with the environment is generative of skills that enlarge the scope of one's capacity and ability. But what about the more intimate forms of knowledge that may emerge through practical work in the mountains? What forms of attachment to landscape and to domesticated animals may arise and be nurtured through this engagement? Animals play a significant role in the emergence of affect generated through engagement with the mountains. Moreover, for herders, work in the mountains is not strictly a matter of physically engaging the material world, but is ultimately generative of ethical reflections. Arguably, Buddhism's favorable disposition toward animals may be underpinned by tradition, as scholarship on Buddhist animal ethics explores. But participation in the "pastoralist landscape" equally contributes to the development, the nurturing, and the sustaining of an ethics of care for animals.

The herder's journey also encompasses a plurality of temporal dimensions. The time of day in which a journey begins dictates how far and how high in the mountains a herder will travel and, accordingly, the places he or she will see on a given day. The cycle of the seasons, in addition, dictates the rhythm of movement through the landscape. A herder's knowledge thus has a manifold temporality. As is evident in their observations of how the physical landscape that surrounds them has changed over time, this temporality spans years and decades. While Ladakhi herders spend almost half the year in the mountains, they rarely venture out to the high pastures in winter, when mobility is hampered by the accumulation of snow. But during the farming season, usually from late spring through autumn, livestock must be kept far away from cropland in order to prevent damage. Traditionally in Sham, from about May to October, the livestock are kept closer to the *phu* (high pasture, in the uppermost part of a valley) in the summer settlement, a place also called *jhonsa* (*bzhos sa*), or "the place of the dairy cattle."[5] Higher in altitude than the villages, these seasonal settlements provide relatively low summer temperatures, ensuring that the milk and its by-products do not spoil. In addition, the vegetation in the *phu* is said to be highly nutritious; Ladakhis maintain that animals will produce good milk throughout the winter if they are well fed in summer.

At the *jhonsa*, people stay in *pu lu*, basic stone shelters, processing the cattle's milk into products such as butter, buttermilk, yoghurt, and dry cheese. The *pu lu* are covered with either tarpaulins and corrugated metal sheets or, in the more traditional style, a canopy made of yak wool. The interior is dark, but the confined spaces are always filled with the sweet aroma of fresh dairy products, which contrasts with the pungent smell of dung outside, as animals are kept in the area at night. The confined space can become quite smoky during cooking time or if the milk is boiled inside, but this has the advantage of chasing away the flies that collect around the *pu lu*. The conditions of life and work at the summer settlement can be quite harsh.

Temperatures can be quite chilly at this altitude, even in summer. Churning butter requires a great deal of physical strength, as is apparent in herders' muscled forearms. Life in the summer settlement also sometimes follows a quasi-nocturnal schedule, with herders waking up in the middle of the night to milk their large herds. Then, as soon as the sun rises, the cattle make their way into the mountains, going as far as the foot of the glaciers, where they spend the day.

As much as the mechanization of agriculture, with tractors and mechanized threshers, renders draft animals redundant, the availability of packaged milk in the market together with a variety of ready-made dairy products such as cheese, yoghurt, and butter, all increasingly popular among Ladakhis, brings into serious question the efforts required to keep herd animals. Yet local dairy products continue to be valued. In the Zanskar region of Ladakh, for example, pastoralist activities still occupy a central role in the local economy. Dairy products from the region, known for their richness, are sought after throughout Ladakh. Part of their uniqueness comes from the animals' special diet of medicinal plants that grow at the foot of Zanskar's glaciers. But according to herders, vegetation in the mountains has become sparse, creating difficulties for dairy herds, which now produce milk of lower quality. Herders frequently associate the decline in vegetation in the pastures with the decreased winter snowfall and the resulting reduction in the size of the glaciers, changing meteorological patterns they link to people's increasing greediness. What defines this cupidity is a lack of care for animals and a consequent abandonment of pastoralist activities, as people are nowadays more interested in "running after money."_

Alternately, in villages that are relatively close to high pastures, animals are kept in corrals within the village and brought up to the grazing land during the day. Otherwise, villagers collectively hire a watchman to do this work. Traditionally, in a system called *ra res*, villagers take turns moving the animals to the pastures, but this system is falling out of use, along with many other community arrangements. This complicates the sustainability of pastoralist work. In any case, whether herders are staying in the summer village or trekking to the mountains daily, movement remains a key feature of herding sheep or keeping an eye on the *dzomos* (female *dzo*) and cows. The only livestock that herders do not need to accompany are the few bulls, which are left to roam free in the high pasture for much of the grazing season. All these activities are part of how the mountains are known and engaged with. Yet this engagement does not happen in isolation; today, herding work is complicated by economic, political, and cultural changes.

TERRITORIAL CLAIMS: OF PREDATORS AND HERDERS

As we drink butter tea with *ama* Youthol in Saspotse, a tiny village bedecked with blooming rhododendrons, our conversation is interrupted by a phone call. The old woman pulls her phone out from her woolen coat (*mo gas*) and receives the bad news: the snow leopard made another attack, this time in Yangthang, a few kilometers away. The big cat has been terrorizing Sham all summer. He attacks not only *dzo* grazing in the summer pastures, but also sheep and goats in their pens, even within the villages (figure 5.1). A few days before, the predator also killed many animals in Skindeyang. Even more shocking is the fact that the usually timid snow leopard has been making his attacks during the day. In Ulley, a tiny village of fewer than ten inhabitants, a man saw the snow leopard on the road and was even able to immortalize the moment with the camera of his cell phone. Apparently, the big cat was not a bit bothered by his presence.

In the hope of meeting herders in the summer pastures, Namgyal and I had embarked on a summer trek through Sham the previous week. From village to village, people recounted the predator's ravages. It seems as though we are following in his tracks—or worse, as Namgyal suggested, perhaps he is following us. As we sit with *ama* Youthol, we reflect that if the snow leopard is consistent, his next attack is going to be here, in Saspotse, or in Likir, within the next few days. That night, barely able to sleep, *ama* Youthol makes several trips between her room and the enclosure next to the house to check on her sheep and goats.

The following morning, as we pack our bags for our trek to Likir, *ama* Youthol generously brings us carefully wrapped roasted chapattis. She seems tired, a state that does not keep her from exhibiting the kindheartedness typical of villagers. I ask her if she is worried for her animals. "It is so difficult these days," she says. Since nobody is there to bring the animals to the pastures, she explains, the poor animals end up spending hours in their

5.1 Villagers looking at sheep and goats killed by a snow leopard that got into their pen. Photo by Jigmet Lhundup.

enclosure, and become "a feast ready to be served to the snow leopard. Old people are left with all the work these days," she laments, shaking her head.

•　•　•

To my great relief, the only wildlife we encounter on the way to Likir are a few Himalayan snowcock. On arriving in the village, we stop for chai next to the monastery, where we meet with Gyatso, one of Namgyal's friends. The building's white walls contrast with the surrounding gray mountains. Decades ago, during the first Indo-Pakistani war, the raiders occupied the monastery, temporarily converting it into a military base. The raiders had summoned the villagers to the monastery, where they were told that nobody should try to escape the village or they would be shot. For emphasis, one man fired a shot toward the mountain, a frightening moment that many still remember. Some people were even held in the rooms of the monastery for a short period of time.

We soon learn from Gyatso that another predator attack took place a few days before: *aba* Stanzin, a local herder, lost one of his *dzomo* to to a pack of *kishang*. As if the big cat's reign of terror was not enough, hungry *kishang* are also stalking villages all over Sham. The canine predators show little fear of humans as they wander into villages and approach houses. In recent years, the situation has become so critical that villagers have grown increasingly wary of the *kishang*. Gyatso confides that he is now afraid for his children's safety when they play outside. His apprehension is well justified. The year prior to my research stay, a colonel was killed by a pack of dogs while out for his early morning run in Leh. The incident was spoken of throughout the region.[6]

In the past, villagers would go into the mountains in search of predator cubs. Out of a litter of wolves, two would be kept alive and the rest killed. This method was also intended to deter adult wolves, as villagers believe that a female wolf will never return to a place where her cubs were killed. Upon bringing the spared cubs back to the village, the hunters would be rewarded with barley grain in a practice called *shang shang*. Displaying the spared cubs before releasing them back into the wild was also a way to prove that the killing had not been indiscriminate. The construction of *shang dong*, or "wolf pits," was another measure to control the population of predators. The *shang dong* is a circular pit with a high wall. To attract the predator, a sheep or a goat is placed at the bottom of the hole, sometimes alive, sometimes dead. Once inside, the curved stone wall prevents a fallen predator from escaping, making it easy to kill. But *shang shang* is no longer practiced in Sham today, nor are *shang dong* any longer in use. Enacted in 1978, the Jammu and Kashmir Wildlife Protection Act banned the killing of wildlife. The act stipulates severe prison sentences for offenders, effectively putting an end to local control mechanisms in Sham.[7]

Knight (2000) points out that disputes with the state over the understanding of wildlife and the sharing of space by humans and animals run through the heart of many human-wildlife conflicts. Moreover, contemporary attempts at wildlife management are typically underpinned by the conceptual segregation of nature and culture (see Igoe and Croucher 2007; Nadasdy 2003).[8] In separating people from their surrounding environment, these measures fail to account for the practical reality of life among predators for Ladakhi herders. Yet if some villagers blame the ban for the increase in wolf and *kishang* attacks, many more believe that the crux of the problem rests in the widespread presence of military installations. The recent increase in the population of *kishang* is widely attributed to easy access to food sources around military compounds and the numerous campsites housing road builders. In addition, it is common practice for the army to deport dogs found wandering around military compounds, and many end up finding their way to the villages.

If the sentence inflicted by the authorities acts as a powerful deterrent to killing animals, the region's growing dominance of Buddhist orthodoxy should also be taken into consideration. While in the past, urial, blue sheep, and goats were sacrificed for religious rituals, such practices are now the object of strong moral disapproval on the part of clergy, contributing to an increased sensitivity regarding the killing of animals. Some feel this also constitutes an additional incentive against transgression of the law.

But the predicament of herders today in Ladakh must also be read against the backdrop of ongoing rural depopulation. The lack of workforce in the villages to bring the animals to pasture, many villagers insist, is the immediate problem. This demographic dynamic cannot be dissociated from the territorial claim the Indian state has been making on Ladakh since the independence of India. The increased militarization of the region, along with an ever-increasing state presence through bureaucratic institutions, in providing employment, has contributed significantly to the marginalization of pastoralist activities. There is a generalized disinterest for herding work in favor of these new employment opportunities (see Bhan 2014:107).

Because elders lack the stamina to climb to the high pastures, they must leave their animals in their pens for extended periods of time. Trapped in an enclosure, the animals become easy prey. Because the aging pastoralist workforce is not being replaced, pastoralism in Sham today increasingly becomes a game in which the odds are stacked against the villagers as they face a multifaceted opponent. In many ways, this is what makes the vicious attack on *aba* Stanzin's *dzomo* by a predator so disheartening: as Gyatso suggests when reporting the incident, "*Aba* Stanzin incarnates a lifestyle that is seemingly disappearing in Ladakh." In Likir, many people speak of *aba* Stanzin as a resilient villager for his determination to continue to trek to the high pasture every day with his herd.

SEARCHING FOR *ABA* STANZIN

Eager to know more about *aba* Stanzin and hear his story, Namgyal and I decide to join him on his daily trek the next day. Interested by our pursuit, Gyatso decides to accompany us. *Aba* Stanzin normally leaves for the *phu* with the early morning light, returning home at dusk. The three of us set off at daybreak for *aba* Stanzin's house, a walk of about an hour. The morning is cold, but the bright sun foretells, it seems, a nice day.

As we reach the upper part of the village, we pass the government-run school where Gyatso's daughter will spend her day. As is the case with many government schools in Ladakh, attendance here is on the wane, as parents increasingly opt to send their children to private schools in Leh and outside Ladakh, which they feel provide a better education. The trend of sending children to boarding schools further contributes to the villages' population drain and threatens the sustainability of government schools. The situation also disconnects the younger generation from rural life.[9] But the impact of schooling on village dynamics is not a new concern. Ladakhi elders recall that when state schooling was first introduced in the late 1950s, some parents hid their children when government officials made the rounds to enforce mandatory enrollment, worried about the lack of working hands to take the animals to pasture.

When we reach *aba* Stanzin's house, his wife, a charming woman of small stature, informs us that he already left for the pasture some time ago with his second wife.[10] We resume our ascension of the valley. The higher we climb, the more rugged the landscape becomes. We hike past terraced fields overgrown with weeds, a common sight throughout Sham. In every village, I meet families, and hear of others, who have stopped cultivating some of their plots of land due to a lack of working hands in their households. As discussed in the previous chapter, there is, for some villagers, a strong sense of shame attached to failing to care for the land and maintain its vitality.

Higher up, we come across a man who is sitting and spinning wool while watching over his sheep and goats grazing in a *spang*. A *spang* is an area of grassy, damp ground fed by underground springs; it is lush with thick grass and moss, ideal for cows, yaks, *dzo*, horses, and donkeys and especially valued for winter grazing. Ladakhi land, which can be defined according to functional, topographical, biological, and geographic criteria, reveals a rich variety that contradicts the untrained eye's first impression of infinite wasteland.[11] Seemingly barren, the mountains contain the precious resources that have for generations sustained the population of Ladakh. Some of the vegetation is the source of medicinal plants for the *amchi*, the practitioners of traditional medicine, while other vitamin-rich plants supplement the local diet.[12] Other resources include shrubs and various woody species, as well as dung produced by grazing animals, precious combustible materials used to heat homes in winter. Even the vast plateau, the *thang*, provides villagers with building materials such as rock and sand, and offers rough grazing for animals. Thus, Ladakh's parched mountains and plateaus abound in resources, and knowledge about their use is passed down through the generations. Before the onset of winter, villagers spend long days in the mountains collecting wild plants and shrubs as fuel. To ensure the sustainability of these resources, village headmen set annual limits on how much each household can collect—though this practice is less needed today, as these resources are increasingly being replaced by cooking fuel and commodities from the market.

We sit, and the herder takes a handful of dried apricots from his pocket, which he shares with us. Namgyal takes the man's drop spindle and unsuccessfully tries to spin the wool. Gyatso tries as well, and fails, as do I. It proves to be much more challenging than it first appears. "*Azhang-le*, why do you think there are more *kishang* these days?" I ask. The man looks at the horizon, locates his herd, and then directs our attention to the nearby stream. On the other side, some fifty meters from us, a herd of urials, Ladakh's wild mountain sheep, are grazing.[13] We all silently observe the animals. Seeing such rare animals up close feels quite fortuitous—but perhaps not entirely. "We see so many of them these days," remarks the man, adding, "these animals have to come all the way down the mountains to search for fresh grass." This behavior, in his view, can only be explained by the fact that the mountains have become very dry due to the reduced snowfall and rainfall. That the mountains have lost their lush green is a common observation among Ladakhis. Like the herder, many believe that this shift contributes to the predation near the villages, as snow leopards and *kishang* follow their prey to lower altitudes.

Namgyal notes that it is time to resume our trek if we want to catch *aba* Stanzin. In order to avoid a marsh, he suggests we walk along the valley's western flank. But soon we come to difficult terrain strewn with boulders. Rust-colored moss covers the boulders in patterns that could have been painted with the brush of an abstract artist. "I think we have reached the beginning of the *phu*," says Gyatso. He is as unfamiliar with this place as we are. Having spent his school years outside, Gyatso never got to explore the pastures of Likir (figure 5.2). "My father spent his life here, grazing the animals for the monastery," he says. "Well," replies Namgyal mockingly, "you could come here in the summer and leisurely make cheese and spin wool instead of changing jobs all the time." Both Namgyal and Gyatso are in the same precarious financial situation. Gyatso has worked different jobs, ranging from trekking guide to porter in the army, and he also unsuccessfully opened and ran a small teashop. "This is more like a hobby for old people who have time for themselves, " replies Gyatso, visibly insulted by Namgyal's suggestion. Gyatso's comment reflects what many people of his generation see in pastoralist work, namely an enterprise that is not lucrative, one done only for the pleasure of being in the mountains and associated with an aging generation. But for those who still attach significance to agro-pastoralist work, it offers an unparalleled work-life balance. As one middle-aged woman told me, "People would be really surprised to know how well you can live by only farming your land and rearing animals." For her, subsistence farming is the key to a good quality of life, instead of always being away "to run after money."

5.2   The *phu* (high pasture) of Likir village

We press on, but progressively bigger boulders cause our pace to slow considerably. Changing direction, we regain easier ground and continue our climb upward. If herders often speak of the pleasures of life in the mountains, they equally stress the physical challenge of bringing the animals to the high pastures. In Ladakh, herding is widely acknowledged to be a difficult trade, requiring as it does navigation of highly uneven, steep, rocky terrain, sometimes in cold temperatures. Indeed, the years of herding take their toll, leading to chronic back and muscle pains. One popular Ladakhi truism holds that herders can be recognized by their slight appearance. Tsewang Dolma, from Nye village, described a typical day of herding:

> In the morning, at sunrise, at about four o'clock, I had 110 goats locked in this pen just beside our house. After we got up, we would water the fields, milk the cows, and do other chores, pack *kholak* [a dough made of roasted barley and butter tea], and go out to graze. I would stay in the mountains until about five, and then we would start gathering the herd back. Then in the evening, around seven, I would get back to the house and lock them back in the pen. All day long, I would walk with the goats and collect dung in the mountains. It was so cold in the mountains, sometimes it was miserable. My feet were red because of the cold. My socks would be completely frozen with ice.

Like most herders, Tsewang Dolma began at a very young age—in her case, at eight years old. She first learned from her uncle, who taught her where to find rich grazing land, which places to avoid and water sources to stay away from because they were poisonous, and how to read the mountains and anticipate the terrain ahead. Good knowledge of the high pastures is indeed vital. This was made clear to me by an old man from Tia who recalled a nearly fatal incident. Once, he said, he was collecting edible plants on a mountain pass along with a group of people, among them a man who was not familiar with the high pastures of the area. Venturing to a spot normally avoided by herders of Tia, the man fell into a crevasse; he was saved only by a stroke of luck, as his collecting bags caught on the ice wall, preventing him from falling further.

Having freed ourselves from the boulders, we reach a woman in her mid-sixties resting on a flat rock. We learn that she is *aba* Stanzin's second wife. Her face is crisscrossed with deep wrinkles from years of grazing animals in the mountains under a piercing sun, and her hands, rough and cracked from work and cold weather, bear the testimony of long years of arduous agrarian and pastoralist work. She is wearing a long khaki-colored woolen coat and traditional Ladakhi woolen shoes. We tell her that we would like to catch up with her husband, but she is not optimistic about our chances. "By now he could be anywhere behind the mountains," she explains, looking at the summits ahead. "I always have a hard time following him, he walks so fast. Now

5.3   Man herding his goats

I am going back home. I would rather help my sister with work than spend my days running after him up here," she says. Had we left earlier, we might have caught up with *aba* Stanzin, but it was becoming increasingly clear that we would spend much of our day searching for him.

Navigating the mountains is the key aspect of a herder's journey. Constant movement shapes their knowledge of the mountains. Herders trek to the pasturelands with their livestock in search of a good grazing spot where they can spend the day. If going with sheep and goats only, a herder covers more ground, since these animals are more nimble. But for herders, traveling the mountains becomes a way of knowing a place. Someone familiar with the high pastures of Likir, for example, would not have become trapped in a cluster of boulders, as we did. And someone familiar with the high pastures of Tia would not have fallen into a crevasse like the poor plant collector did.

Animals play a central role in how men and women develop knowledge in the mountains. They dictate patterns of mobility as they choose one path or another in search of good feed. The herder controls their movements to a certain extent, but it is the animals that determine the overall direction. In the words of one herder, "There are always some that lead, and some that lag behind. Those that lead decide the way." While domestication is normally seen as the control of humans over animals, Donna Haraway (2008) invites us to think of domestication in a different way. Humans and domesticated animals, she suggests, have evolved conjointly, in a complex and asymmetrical relation of dependency and interdependence. She uses the trope of "companion species" to deconstruct the assumed dividing line between human and animal. If humans shape companion species, the opposite is equally true. In Ladakh, a herder's relationship with his or her animals channels knowledge about the high pastures.[14] In this way, they are central to the constitution of the herder, playing an active role in how people know a place and acquire knowledge about their environment. As we continue our climb, the craggy mountains surrounding the valley become more distant, and a vast grassland opens before us. However, there is no sign of human presence; our chances of catching up with *aba* Stanzin seem to be diminishing. We nevertheless decide to persevere, but only after a welcome lunch break, as we share the chapatti, eggs, and boiled potatoes that Gyatso has packed. Ahead of us is the Ladakh range, a stunning skyline of snow-covered summits, beyond which lies the Nubra Valley. Crossing this high mountain range to visit relatives would have been common in the past. This is also where, decades ago, Hari Chand escaped toward the Nubra Valley after he, along with his commando, destroyed the howitzer in Basgo village during the war. Behind us, the peaks of the Zanskar range reach into a compact blanket of clouds. Nearby, a lammergeyer, a bearded vulture, is circling, moving its wings with ease. A scavenger species that lives on a diet consisting almost strictly of bone marrow, the lammergeyer of Ladakh are probably quite content with the increasing number of snow leopard and *kishang* attacks. "Perhaps it is looking at the remains of *aba* Stanzin's *dzomo* somewhere ahead," suggests Namgyal. We decide to walk in the bird's direction.

After some time, we come upon a group of abandoned and deteriorating *pu lu*. The scene speaks for itself. The *pu lu* are vestiges of an era that is rapidly coming to a close in Sham today. A bit more than a decade ago, this place would have been alive with herders and animals. But now the *pu lu* is no more than a handful of deserted stone shelters. There is no one to be seen, nobody to put the fallen stones back into the walls. The deserted summer settlement stands in testimony to a disappearing way of life and body of knowledge.

Suddenly Namgyal, who is walking ahead, calls to us. On the ground, next to a large rock, are the remains of a recently killed animal, likely *aba* Stanzin's *dzomo* (figure 5.4). It is clear that the beast did not stand a chance. Looking at the carcass, we all stand amazed at the violence it bespeaks.

5.4   Namgyal standing near the remains of a recently killed animal,
        likely the *dzomo* belonging to *aba* Stanzin

At this point, the sky has filled with ominous clouds and a cold wind is
blowing. We have ascended about 800 meters, at a leisurely pace that has
taken us many hours. It is well into the afternoon and the village below
seems so far away. *Aba* Stanzin is somewhere in the upper part of the valley,
probably very far ahead. We agree to abandon our search.

The trek to the *phu* on which Namgyal, Gyatso, and I embarked in Likir encapsulates the essence of a changing pastoralist landscape. Instead of finding a place occupied by herders, we found dilapidated seasonal settlements, uncultivated farmlands, empty grazing grounds above tree line, and a near absence of human presence. The scenario would have been quite different a decade ago, when the high pastures of Sham played a key role in local economic activities.[15]

But the predicament of herders in Ladakh also stems from a changing attitude toward the presence of animals in villages. Time-honored tradition holds that animals be allowed to roam free in the villages after harvesting, but in one village after the other I heard recurring disagreement over when this should take place. As discussed in chapter 4, agrarian activities within a village are no longer done in synchronicity; as a result, not every household completes its harvest at the same time. In the absence of a clear consensus among villagers over when to let animals graze freely, tensions build. The new lucrative business of tree plantations is an additional complication for herders, since goats often eat the bark; as a result, many villagers see them as an irritating presence.

Yet if some consider animals a nuisance, many others see their presence as a blessing, for it brings liveliness to a place. These people speak of their animals with great affection. "You should see how intelligent they are," one man told me, referring to his goats. "They are like my small children. How could I get rid of them?" These idioms of filiation are emblematic of Ladakhi herders' relationship to their domesticated animals.

An incident that took place during my stay in Hemis Shukpachan exemplifies the bond between humans and domesticated animals in Ladakh, and its transcendence of species. When Dolma, the mother of my host family, left early in the morning for a wedding ceremony in Alchi, it did not take long for the family's only *dzomo* to break the gate of her corral and run away. Panicked, Dolma's husband looked in vain for the animal. Wandering alone in the mountains, the *dzomo* would be an easy prey to dogs and wolves. When Dolma came back home two days later and learned about the *dzomo*, she was obviously worried. "She does that when I go," she said, observing that she only rarely left home, but every time she did, the *dzomo* would run away. "She is upset and she is looking for me. She cannot be separated from her mother." The *dzomo* eventually made it back to her corral in the middle of the night. "She knew I was back. She felt it," explained Dolma, smiling.

Ladakhis indeed frequently refer to their domesticated animals using idioms that convey a sense of "interspecies kinship" (Govindrajan 2015a:507–8). This mode of engagement evokes personalized and meaningful relationships that are constituted through communication, as when Dolma's *dzomo*, in despair, escaped to find her, or through affectionate gestures, as when the man played with his baby goat like he would play with his own child.

These idioms of kinship contrast markedly with the idea of domesticated animals as a nuisance, which increasingly frames how these animals are considered. This signals that the pastoralist landscape of Ladakh is currently changing, as the relations that make this landscape are redefined by a changing attitude toward animals. For some, animals have become a burden, not only because they can damage trees and crops, but also because they must be fed, either by being brought to pasture or by being fed in their corral. This requires extensive work, which may be viewed negatively if not anchored in meaningful relationships and a sense of responsibility toward the care of others. Structural constraints currently contribute to making animals increasingly anonymous, thus rendering these relations less meaningful.

In his study of Cree hunting, anthropologist Paul Nadasdy explains that considering local people's view on human-animal relationships implies rethinking notions of personhood and agency. For the Cree hunter, animals are more than a symbolic construct: thus, hunting, rather than simply an economic affair that hinges on an act of violence, is "a process of reciprocal exchange between hunters and other-than-human persons" (Nadasdy 2007:26).[16] In the same way, Ladakhis, acknowledging that domesticated animals have long been essential to human life in the harsh environment of Ladakh, locate reciprocity between humans and animals at the heart of pastoralism. Putting an end to pastoralist activities is nothing less, in their words, than "abandoning" these animals. As much as taking care of the land is a duty, taking care of the animals is a moral responsibility. For this reason, the decreased interest in pastoralist activities is viewed by many as a moral failing.

The care of domesticated animals is a source of concern for many elders today, one also associated with the Tibetan Buddhist conception of life cycle. As Geoff Childs (2004:131) and Thubten Jigme Norbu and Colin Turnbull (1972:75) observe, for Tibetan Buddhists old age is a stage when attitudes toward animals change: as one prepares a future reincarnation, no negative treatment should be inflicted on animals. Because the lot of sold animals is likely slaughter, moral dilemma is also one facet of the predicament herders face in Ladakh. This is quite tangible in the following reflection by Nawang Zangpo, an elderly man from Likir:

Now the children no longer stay at home and the old people are not able to take care of everything alone, so there is no choice but to sell off the animals. The Balti come here and they buy all the cattle that the villagers don't want. Some people say the Balti are taking these animals to kill and eat them! So when they come, we ask them if they are going to kill the animals, and if so, then we tell them that we are not ready to sell. They say they are not going to kill the animals, that they will rear them. But they say that only to make us happy; god knows what they do with the animals.

Today, as the younger generation becomes less likely to step in to replace an aging pastoralist workforce, many elders are faced with the extremely difficult decision of selling their cattle, a decision fraught with moral ambiguity given the likely fate of these animals. For the reasons expressed by Nawang Zangpo, many Ladakhis see in the younger generation's disengagement from pastoralist activities one more step in the decline of moral values. But elders are also inclined to blame their own physical incapacity for the sad lot of domesticated animals. Herding is physically strenuous. Days spent in the high pastures, especially with sheep and goats, which are very mobile, are particularly challenging and become increasingly difficult with old age. As one old man told me, "It is only because I am old that I cannot go in the mountains anymore with the animals." For him, being a herder was a way of life that could only be hindered by the limits of his own body.

But herders' ethics of care must also be considered in light of the fact that domesticated animals are seen to contribute to the fortune of humans. Elders' correlation of the decreased presence of animals with the degradation of the environment was too often postulated to be ignored. According to Tibetan Buddhist cosmology, everything is interconnected, and human deeds and caring have an effect on the environment. Reflecting on the causes of glacier recession, an elderly woman who spent her life in the mountains with the animals told me, "People are unfortunate nowadays. In those days, the people were fortunate, even though they did not have a good education. *Konjok*, they were fortunate because they had cattle."

CONCLUSION: A LANDSCAPE OF PREDICAMENTS

The pastoralist landscape of Ladakh is based on reciprocal relations between humans and domesticated animals. It is a collective entity that emerges from a network of practices that sustain pastoralist activities. These practices include community arrangements, the provision of a workforce to go into the mountains, the care given to cattle, the mechanisms that allow predator control, the relocation to the summer pasture to process dairy products, and the integration of these products within the household economy. But many of these practices are today being abandoned or restricted, and as a result, for those who still strive to maintain herding activities, the pastoralist landscape is increasingly becoming a landscape of predicaments.

Participation in the pastoralist landscape generates knowledge through mobility in the mountains. This knowledge hinges on a process of embodiment through physical labor and sensory perceptions. It also has a temporal rhythm, as it develops during a specific period of the year and spans many years. Animals are central to the production of this knowledge, as it is through their movement to the high pastures that herders become familiar with the mountains. But engagement with the mountains does not only enable knowing a place and navigating the pastures. It also fosters attachment to domesticated animals. As elders insist, pastoralist work is more than a mere economic activity. At the heart of herding are relations of care and reciprocity between animals and humans: animals sustain human life; therefore, humans must sustain animal life in turn.

The relations that constitute the pastoralist landscape are neither static nor isolated but are shaped by changing cultural practices and a redefined economy. The contemporary economic structures of Ladakh contribute to the disassembling of the pastoralist landscape. In reducing patterns of mobility in the mountains, the restructured local economy has implications both for the way the landscape is known and for meaningful engagement with animals. Affection for animals—or, conversely, antipathy toward their presence—may develop through interactions. These interactions can be favorable, as when a herder is amazed by his goat's intelligence. But they can also be irritating, as when a villager's crop is damaged by a neighbor's goat. But the relations between humans and other species are reflective of relations among humans and, as scholars have demonstrated, can also be influenced by a broader political context (Govindrajan 2015b; Jacobs 2001).

The predicament of herders is therefore multifaceted and also profoundly moral. The combined pressure exerted on pastoralist work has resulted in the increasing sale of domesticated animals, a trend that is especially upsetting to elders, whose emotional and moral engagement with their animals has been a defining fact of life and work in the mountains. In the view of many elders, the failure to care for these animals suggests a changing moral order that goes hand in hand with environmental change, for the human and natural realms are inseparable and evolve conjointly. Moreover, just as elders feel that they have become a burden for younger generations, they see their domesticated animals facing the same lot. But the disentanglement of the pastoralist landscape also raises questions for an ethics of care for animals. The absence of meaningful relationships may contribute to rendering these domesticated animals nameless, such that their presence is seen as unpleasant and contentious. This suggests that meaningful interactions with domesticated animals contribute to the development of an ethics of care, the nurturing of which is today increasingly curtailed.

CHAPTER SIX

Intimate Glaciers and an Ethics of Care
Mutual Recessions

For landscapes are always available to their seasoned inhabitants in
more than material terms.

—KEITH BASSO, *WISDOM SITS IN PLACES*

An old *dzo* may have no strength
But he knows the road better

—TIBETAN PROVERB

"YOUNG PEOPLE DON'T KNOW ABOUT THE GLACIERS TODAY, AS MANY
of them have never reached the *phu* [high pasture]," Dolma tells me. She is
trying to convince me that renting her horses and trekking to the glacier with
her husband, Puntsok, a man in his late sixties, is going to be a much more
valuable experience than going with any village youth, as they "know nothing
about this place." Namgyal and I are in her native Hemis Shukpachan village.
Dolma, her mother, and her granddaughter are shelling peas on the sun-
warmed terrace next to their house. Each woman contributes to the melodic
rhythm of the peas landing in the plastic pot: Dolma's steady cadence is punctuated by the slower movements of her mother's arthritis-afflicted fingers and
the sporadic participation of her granddaughter, still learning. The scene
evokes skills passed down through generations, honed through practice but
eventually hindered by the effects of a lifetime of hard work.

My plan to trek to Shali Kangri, the local glacier, is motivated by a desire
to see the alpine grassland that many of my elderly interlocutors passionately
describe while recalling their days as herders in the mountains. Although

139

their bodies no longer have the stamina to trek to the high pastures, many still long for the place where they once spent half of each year. I also want to bring back photographs of Shali Kangri. In the village, many elders have remarked how they saw the glacier recede during their life, but the last time most of them could see Shali Kangri was more than a decade ago. In Sham glaciers tend to be visible only from the high pastures, a difficult trek for aging men and women. By seeing pictures of the glacier, they can assess any further changes that have occurred. This is what I am trying to arrange with Dolma and Puntsok.[1]

As is typical in a Ladakhi village, passers-by join us, slowly increasing the size of our group; people chat and make jokes. An old man tries to warn me against potential disappointment: "The glacier is not as it was; it is really rather small today and there is nothing much to see there." This shifts the discussion to the question of melting glaciers in Ladakh generally. By then twenty-year-old Stanzin has joined our group. His outfit, clean jeans and a slim-fitting jersey, contrasts with the elders' traditional long woolen jackets. Like many young Ladakhis, for the past several years Stanzin has been residing outside Ladakh, first to attend college, and now he has enrolled in a university program. He is visiting his home village during a study break and seems quite eager to reconnect with village life. At one point, Stanzin interjects: "These changes are due to global warming." While most younger Ladakhis are familiar with the scientific interpretation of climate change, elders have their own distinct view of these phenomena. "What are you talking about?" asks Dolma. "All this is occurring because people have become bad these days." Stanzin, obviously perplexed by this comment, remains silent. "Look at what people are doing today," Dolma continues, "always running after money and not taking care of things. People don't have the time to go to the mountains with the animals these days, so they don't care about the glaciers. To care for the glacier, you have to see the glacier, you have to know the glacier, like you know a friend. People have become careless, unmindful, and that is why there is no more snow," she laments. "The glaciers are going because of that." "Nobody cares," says another woman, as if completing Dolma's thought. "People have become blind today."

Dolma's remarks echo the views expressed by the vast majority of my elderly Buddhist interlocutors. The perception that today's glacier recession is taking place in a moral vacuum can be traced to the Tibetan Buddhist principle of reciprocity as an ideal of harmonious cooperation, manifest in humanity's relationship with the natural world (P. Harvey 2000). Many Buddhist Ladakhis believe that there is an entanglement between humans and glaciers that translates into a reflexivity between people's moral attitude and the state of the glaciers. Very often, when I asked my interlocutors why the glaciers are receding, they would answer laconically: "Because people have become bad."

In Ladakh, glaciers are constituted as the object of morality and subject of care through labor practices. According to local beliefs, glaciers reflect a moral order, and their decaying state signals a failure of proper social conduct. This belief is not unique to Ladakh and indeed prevails in Tibetan Buddhist areas (Diemberger 2013:102–3).[2] The decaying morality to which Ladakhi elders refer, as examined in the previous chapters, has a long genealogy in Ladakh and is associated with the political and economic processes that transformed Ladakh as a place after the independence of India. For some, it is rooted in a new epoch, set by the first Indo-Pakistani war, an event that, with all its violence, left open wounds and transformed people permanently. For others, the decaying morality cannot be divorced from the decentering of the agro-pastoralist economy and access to employment outside the villages, resulting in village depopulation. It is thus a symptom of a changing attitude, one that leads people to abandon their community responsibility to embrace paid work, leaving the land and animals in the care of increasingly isolated villagers. It is against this backdrop that comments on the pecuniary enthusiasm of people today take on full significance. The idea of reflexivity between human actions and the condition of glaciers has acquired a deeper meaning in the past decade, as the impacts of climate change grow more apparent. For many Ladakhis, the recession of glaciers is a consequence of the changing moral values brought by economic changes in the region, a rationalization based in beliefs about the entanglement of humans with glaciers and the connection between attitudes and the state of the environment.

But what might nurture a sense of care for these ice summits? In linking a lack of knowledge about glaciers and a certain apathy regarding their recession, Dolma suggests that caring for glaciers requires being in contact with them. The glaciers to which she refers are the village glaciers, which in Sham are often located deep in the valleys above the settlements; their visibility therefore requires an engagement with the mountains.[3] Shammas tend to have a strong connection with the main glacier of their village, to which they often refer by a distinct local name and the possessive "our"—something they do not do when speaking of the glaciers of other villages.[4] These glaciers, being at relatively low altitudes, are accessible, but only by physically going into the mountains. Significantly, when Dolma laments that the younger generation is not familiar with glaciers because they have rarely, if ever, ventured out to the community's high pasture, she, like many other elders, may be oblivious to the fact that younger Ladakhis often become familiar with glaciers through their work, for instance as tour guides. But it is intimate knowledge about glaciers more than superficial familiarity that concerns her. She regrets the lack of connection and sense of care that people develop with a village's main glacier, something that can be cultivated only through sustained engagement with the mountains. In her view, this form of embodied knowledge and practice is central to the fostering of an ethics of care.

In Sham, through labor in the mountains, landscapes and the elements that form them are brought into a network of relations that comprises notions of morality, imaginaries, and affect, which together lead to the development of an ethics of care for glaciers. For if glaciers in Ladakh are, as in many parts of the world, barometers of morality (Gagné, Rasmussen, and Orlove 2014), as discussed in the introduction to this book, they are equally elements of the natural landscape that need to be attended to. Shifting the long-held focus on ethics as a collective and social affair, scholars have recently departed from a symbolic understanding of religion, emphasizing instead individual perspectives and giving centrality to the experiential dimensions of ethics. Although religion has remained a central concern in many of these writings (Asad 1993; Mahmood 2005; Laidlaw 2014), others have taken a different avenue, analyzing ethics through the prism of ordinary language (Das 2006; Lambek 2010) or exploring how ethical issues emerge in the mundane course of everyday life (Stafford 2013). However, ethical dispositions toward the environment, their practical dimensions and how they develop among moral subjects, remain largely unexamined. Conversely, while scripture and doctrine inform us of how Buddhism values the environment, reliance on text provides only a partial window into ethical dispositions, often failing to account for how these may develop and be nurtured through practical experience of the world.

Through work in the mountains and the experience of the high pastures, herders of Ladakh develop knowledge about glaciers. This awareness forms through sociality, encounters with a realm of beliefs, and the embodied experience of the mountains, which together produce an intimate knowledge that contributes, for many, to attachment and a sense of care for glaciers. As Dolma suggests, caring for glaciers, or listening to what they tell and what they ask of people, requires knowing about them.

PRODUCING INTIMATE GLACIERS

In his autobiography, Thubten Jigme Norbu, brother of the current Dalai Lama, describes the deep attachment he felt as a child for Kyeri, the glacier overlooking his natal village in Tibet, visible from his home:

Kyeri means something like "Mountain of Happiness," and for me it has certainly become the embodiment of a happy childhood. It was under Kyeri that I experienced the love of my parents and my brother and sister, that I rode our horses over the flower-dotted pastures, and accompanied the herdsman and his beasts. It was towards Kyeri that our prayers were directed, because Kyeri was the throne of our protective deity and bore his name. Throughout life the idea of terrestrial and heavenly happiness will always be connected with Kyeri. For me Kyeri has become the symbol of a full life. (Norbu and Harrer 1986:49)

The feelings expressed by Thubten Jigme Norbu are similar to those evoked by some of the Ladakhi herders I met.[5] Their memories communicate an intense physical and sensory experience, as exemplified by the recollections of a ninety-year-old man who told me about a glacier's magnetic effect:

I used to leave the *dzo* to graze freely in the mountains, and I would take salt for them twice in the summer. One time, the *dzo* had reached the foot of the glacier, closer than usual. I gave them the salt and drank the *chang* [homemade barley beer] I had with me. I looked at the glacier and I thought of going closer. But I was alone. I finally decided to go. There was first a mound, and I climbed further on. Then I could see the glacier even better, and it was really beautiful. I walked on another mound and reached the glacier. I could have gone for a circumambulation [around the mountain]. But on to the other side, the whole valley was completely packed with ice. The glacier was immense. I was alone, so I decided not to go.

These accounts suggest that a feeling of deep connection may be a common aspect of glacier engagement across different settings and contexts. However, what makes the experience of herders in Sham different from that of Thubten Jigme Norbu is the centrality of time and movement in how herders come to know the glaciers. As seen in the previous chapter, time and movement are defining aspects of a herder's engagement with the mountains. Here, I wish to emphasize how important these dimensions are in the production of intimate glaciers, such that, as *ani* Dolma suggests, one comes to know a glacier as one would a friend and, consequently, to care for its condition. In general in Sham, village glaciers are not visible directly from people's homes, as they were for Thubten Jigme Norbu. In Sham, the sight of these glaciers need not involve a hazardous trek or occur only once in a lifetime. Rather, seeing the glacier of a village entails engagement with the high pastures in the course of seasonal labor, repeatedly over the years.

Insights into both place and knowledge about place have led to an increased interest in place-making activities and the interactive aspect of situated knowledge. Focusing on work practices, scholars have illuminated how embodied interaction with the material world generates new skills and knowledge (Ingold 2000; Marchand 2010). But as Laura Bear (2014:72) points out, interaction with the material world generates affective experiences and ethical reflections as well. A consideration for such feelings and sentiments can shed light on how herders nurture attachment to glaciers through their work.

Affect has been described as an intense, interactive bodily experience, "the feeling of having a feeling, a potential that emerges in the gap between movement and rest," the experience of having "unconscious responses to stimuli and potential perceptions that a subject may or may not perceive" (Rutherford 2016:286). In focusing on the intensities that arise at the convergence of materiality and subjectivity during encounters between humans and nonhumans, affect theory draws attention to the beliefs with which the material world is invested. This constitutes a salient landscape feature in parts of the Himalayas, the mountains being envisioned as the dwelling-place of deities. These deities are often conceived of as transcending space, but they are also said to dwell in specific places, such as in features of the natural landscape.[6] These beliefs are reflected in systematic conceptions of space and place—such as Buddhist ideas about verticality, which hold that heights are associated with the gods and purity (Dollfus 1989:131–33; Mac-Donald 1997; Vanquaille and Vets 1998). The sacred character of places is delineated through practice as well, whether sanctified through circumambulation (Buffetrille 1998) or by marking, by means of symbolic architecture, the threshold between inhabited and uninhabited spaces (Dollfus 1989:107; Dollfus and Labbal 2009b:85; Michaels 1998:16).

Overall, this scholarship informs us of the broader conceptual realm through which places in the Buddhist Himalayas are invested with meaning, whether reflected in rituals or the built environment. However, the high pasture as a place has received little attention, especially in Ladakh where it constitutes a key site for engagement with glaciers. Focusing on the work of herders illuminates how this realm of beliefs is encountered while grazing animals in the mountains. In Sham, although none of the village glaciers were described to me as sacred (*gnas ris*), and although a high pasture is not considered sacred unto itself, the way to a glacier is punctuated by the presence of spirits of different kinds and elements of a sacred nature, such as the caves where renowned saints are said to have retreated, or sacred lakes.[7] Moreover, in the high pastures, moral narratives are associated with elements of the natural landscape.[8] But how is this realm of meaning encountered outside a ritual context, through mundane labor?

Focusing on the experiential dimension of work in the mountains reveals that engaging with the high pasture is not only about entering a realm of beliefs, but also about generating new forms of knowledge. Following the work of Donna Haraway (1991, 2008) and her overcoming of the nature/culture divide in favor of a processual understanding of life, we might want to think of how herders are "becoming" *with* glaciers in a process of ongoing relationality. The pasture is an assemblage of collective species that becomes a "contact zone" (Haraway 2008:205), and as herders engage in pastoralist life processes, they are shaped by the mountains and develop knowledge about them and their glaciers. This leads them to develop affective attachment and an awareness for glaciers and their state. This process of learning and becoming, by which humans are not distinct from or masters over nature, and through which they develop an ethics of care for glaciers, is today increasingly curtailed by political and economic processes that drive people away from the high pastures and, therefore, from glaciers. In this regard, village glaciers are increasingly becoming an abstraction, an element of the landscape that people know less and less from first-hand experience. But for those who have long engaged with the mountains, glaciers emerge from imaginings about a sacred geography, from the materiality of the mountains, and, as we saw in chapter 5, from an engagement with the multiple relational processes between humans and nonhumans that take place in the high pastures. Knowledge about glaciers is therefore profoundly intimate and "situated." It is precisely the belief that glaciers are the product of local variables that leads Dolma to shun an idea as exogenous as climate change to explain their state. In fact, receding glaciers are hallmarks of climate change, but knowledge about them tends to be expressed in a normative language that has little resonance locally (see Marin and Berkes 2013). This knowledge, in its objectified nature, fails to account for the fact that glaciers and ice bodies are not only infused with cultural meaning, but also known through sensory engagement.[9]

In Ladakh, knowledge about glaciers is contingent on the mobilization of the senses, as well as on the realm of beliefs that inhabits herders on their journey to the high pastures. As I discuss below, herders speak of their engagement with glaciers in a way that foregrounds feelings and emotions. Sensory engagement, in combination with the beliefs with which the mountains are infused, nurtures these feelings of connection with glaciers. Hugh Raffles (2002:332), reflecting on typologies of knowledge about nature, calls on us to rethink the local "as a site of intimacy," broadly defined as the domain of affect. Lacking universalism and thus inaccessible to outsiders, intimacy-as-affect foregrounds the temporal and spatial dimensions of encounters between humans and nonhumans and the embodied practices that characterize those encounters. In emphasizing feelings and sensations, Raffles's perspective allows us to think of the high pastures of Ladakh as sites for the shaping of an embodied knowledge that arises at the intersection between the materiality of a place and broader ideas about the natural world. Moreover, the notion of intimacy helps us make sense of how people develop an attachment to glaciers and an ethics of care through work in the mountains. For as scholars have demonstrated, interaction with landscapes may be generative of affective dispositions (Basso 1996; Haberman 2013; Gold 2001) that in turn may cultivate ethical inclinations (Pandian and Ali 2010:10).

CARE AND THE TEMPORALITY OF GLACIERS

The temporality of glaciers is a key dimension of the intimacy that herders develop in their interactions with them: years spent in the high pastures bring herders into affective proximity to the ice summits, enabling them to decode their physical changes over time. A woman named Tsewang Lhamo, one of the last herders of Nye to regularly go to the high pastures with her animals, told me about the feelings of wonder she felt for the glacier of her natal village and her sadness when she realized how much it has receded over the years:

> One day, I reached very close to the glacier, to its foot. I was with the animals and carrying my child on my back and two heavy bags [full of dung]. The glacier was making a loud rumbling noise. The sound was so loud that when I approached closer to the glacier through the steep narrow gorge, I could hear it rumbling even before I saw its snow-capped peak. It made me shiver; I was a bit afraid. This glacier had a big ice cave at its foot. Perhaps it was the sound of the water coming from below the glacier, echoing in the cave. I always wondered. It was beautiful, just below the glacier; the ground was full of flowers.

Tsewang Lhamo's awe is just as tangible as that expressed by the nonagenarian above, and indicates a profoundly embodied way of knowing glaciers.

In both cases, the engagement with the physicality of the mountains and the beliefs about the place are profound bodily experiences that lead to the emergence of what Massumi (2002:24) calls an embodied "intensity." This intensity emerges from the physical exertion of traversing the mountains and from the beliefs and narratives that infuse the high pastures with meaning. These beliefs contribute to the sublime character of the glacier, but also to the feeling of fear it generates, equally palpable in both herders' accounts. For the high mountains underscore a person's vulnerability, not just due to physical exertion, but also because of potential attacks from malevolent beings. As we saw in chapter 5, mountains are said to be inhabited by cruel *smanmo* (female ghosts) who can abduct people.[10] The high mountains also harbor deities, and as Buddhist Ladakhis know, it is important not to irritate them through any form of misconduct, as this could result in retributions.

But these beliefs and feelings do not preclude objective observations, which are equally constitutive of intimate glaciers. Herders like Tsewang Lhamo often find themselves near glaciers while grazing their animals, which allows them to observe the glaciers' condition over time. Tsewang Lhamo had been trekking to the high pasture with animals from the age of eight. A few years ago, she had seen with great dismay that much of her village glacier had melted, ice cover giving way to bare rocks. The cave was no more, the flowers had all vanished. Even more striking was the glacier's silence: its local name, Chilchil, an onomatopoeic name that imitates its loud rumble, had lost its meaning. "Had there been snow in the winter, *our* glacier would be as it was. Seeing this made me sad," she confided.

In fact, Tsewang Lhamo's sense of despair upon seeing the decaying state of the glacier was so profound that she mobilized her fellow villagers to address the problem. Together they requested the help of a monk who performed a religious ritual, which included prayers and the burying of a sacred vase in the valley that leads to the summit. This was their way of attending to the glacier, an intervention that is part of the conceived entanglement among glaciers, invisible forces, and villagers, in mutual relations of care and reciprocity.[11] For Tsewang Lhamo, it was clear that the melting of the glacier was a direct result of the desecration of this entanglement. These beliefs and recurring contact are determinant for the role of glaciers in a broader affective context.

Like the narration of landscapes or the act of tilling the soil observed in different contexts (Basso 1996; Pandian 2009), the performance of rituals such as the one initiated by Tsewang Lhamo constitutes an act of dwelling, revealing the ethical connection that people have with places and the various forms of beings that constitute them (see de la Cadena 2015). More significantly for a discussion on the shaping of an ethics of care, Tsewang Lhamo's account shows that glacier engagement in Ladakh is not reducible to the feelings and sensations it entails nor to the factual monitoring of changes: as more than a subjective domain of experience, that engagement contributes to the development of a sense of self and a sense of place. In this way, the intimacy of glaciers hinges on a wealth of embodied experiences and beliefs. The engagement over time with the material and affective environment of the high pastures, where herders perform strenuous work, experience intense feelings, develop acute sensory perception, and meet a world of morality, is central to how glaciers are known. To illustrate this, I turn to my own experience of trekking to Shali Kangri. If perhaps self-centered, this account is productive in shedding some light on how practical engagement with the mountains is constitutive of the high pastures as a site of alterity.

IN THE REALM OF SHALI KANGRI

"Why would a young fit man like you be incapable of reaching Shali Kangri?" *meme* Tashi asks Namgyal. The old man is sitting next to an oversized prayer wheel, his usual resting place during his afternoon walk around the village. He wears a worn-out white knit hat and a heavy blue jacket, despite the warm summer weather. "I think you are just being lazy," he continues, before bursting into a long, heartfelt bout of laughter that reveals his toothless mouth. Namgyal laughs at the old man's suggestion too, but with obvious discomfort. "Look at me," *meme* Tashi continues. "I am old and crippled, beyond eighty years old, but until a few summers ago I would still go in the mountains."

When I mentioned my idea of going to Shali Kangri some weeks ago, Namgyal insisted on helping me plan the trip, but he made it clear he would not accompany me onto the glacier. Any ascension above 4,500 meters has an unsettling physical effect on him, causing him to panic. In his previous career as a trekking guide, every time he embarked on a new trek, the anticipation of this sensation would nurture a growing fear. The situation had become so unbearable that he'd decided to retire from this work.

I had learned about Namgyal's fear when we attempted to reach the pastures of Nyemo village a few months earlier. As we were climbing the long narrow gorge behind the village, Namgyal, who had been walking at a steady pace and in silence, was suddenly overcome by a panic attack. He sat down, explaining that he was unwell and that we had to go back. I callously blamed his failure on his overindulgence in *chang* the night before. But Namgyal was truly in distress. "Maybe I don't deserve it," he said, trying to make sense of his failure. "Maybe I am not good enough to be able to climb the mountains." The idea that one needs to be meritorious to be able to reach the high mountains calls on the Buddhist notion of verticality, implying purity and sacredness. But more personally, it strikes at Namgyal's own sense of self. He once confided that being unable to keep a job or to meet the expectations that come with the quality of the education he received left him feeling like an undeserving person.

It is his desire to be a good person that has compelled him to bring me to *meme* Tashi this afternoon. *Meme* Tashi, a former herder, was the first person, early on in my fieldwork, to call my attention to people's changing relationship with glaciers in Ladakh. Since the beginning of my fieldwork, he has been a regular interlocutor, one Namgyal grew particularly close to. Namgyal is worried about my capacity to handle the trek, and we are meeting *meme* Tashi this afternoon to hear his thoughts on doing the trek in just one day. The glacier is some 12 kilometers from the village—just close enough to make the round trip before sunset. Although this is a feasible undertaking, during the past few days several villagers have cautioned me to consider stopping half-way for the night to make the journey easier. However, recent attacks by dogs and snow leopards on cattle in the area have left me leery of sleeping in the open. The foot of the glacier is an ascent of roughly 1,500 meters, and altogether we will attain an altitude of slightly above 5,000 meters. Many tourists accomplish such climbs in various parts of Ladakh, even without elaborate preparation, and given that I have trekked more than once to high altitudes during my stay and that we will ride horses, I reason that the excursion will be easy. *Meme* Tashi suggests caution, however. "It is possible to do it in one day, but you can also suffer from *lapotse* (*la pog ches*), so perhaps you should sleep on the way," he says.[12] *Lapotse* is the Ladakhi term for acute mountain sickness, consisting of a severe headache and a feeling of confusion; it can lead, in the most extreme cases, to cerebral and pulmonary edema. Even if a herder's endurance develops over the years, one should nonetheless always remain alert to any physical discomfort, especially when climbing high in the mountains.

"It looks like this is what happens to you every time you climb, *nono*," *meme* Tashi says to Namgyal, who replies jokingly with his personal excuse, which weighs on him more than he cares to admit: "Maybe I don't deserve to climb the mountains." "Then you'd better start being a good man before you get old like me," *meme* Tashi says mockingly, "otherwise you will end up like me, having to spend your days spinning the prayer wheel!" Namgyal bursts into laughter. Amused, he asks, "What have you done that is so bad in your life, *meme-le*?" "Many things, *nono*," *meme* Tashi replies in a more serious tone. "Many things not worth sharing."

Turning to me, *meme* Tashi continues: "When you climb to Shali Kangri, you will see Mi Duk Sa and Zangs Lung Nge Go from closer." He points at two rocky peaks in the distant mountains behind us. "During the war, the Indian army and the Pakistani army were standing on each of these peaks. From there, they would shoot at each other, and we could hear that from the village." Over the past few months, we had been discussing the war extensively with *meme* Tashi. "If you go there, you will see that there are still bunkers," he says. "In that mountain where the Pakistanis were hiding, they [the villagers] say there are *smanmos*." The war and the suffering it brought have forever transformed the mountains that bound Hemis Shukpachan. Its specter continues to haunt memories and narratives.

"At that time, I had to stay to patrol the mountains with a gun for the Indian army," explains *meme* Tashi, who would have preferred hiding in the mountains and not getting involved in the conflict, like many of his fellow villagers. But having been conscripted, he had no choice. "I was young and innocent at the time. I did not know anything. I made mistakes. I was not of good help to our Indian army," he confides. "Later on, when the Pakistanis were chased away, they did not want me in the [Indian] army. After the war, many stayed in the army for the money, but I would not have stayed anyway." Instead, *meme* Tashi started to go to the high pastures with his animals and those of other villagers who were now missing working hands as many men went into the army. "I spent many years going into the mountains with the animals. In this way I came to understand many things. Then I did some work for the monastery, cleaning, taking care of things." He pauses and spins his prayer wheel. "And now I spend my days praying."

"It is very easy these days for young men like you," *meme* Tashi continues, looking at Namgyal. "You just have to show tourists around and they give you money," he says, bursting into laughter. "We had much tougher times in those days. You children today are spoiled." Namgyal chuckles, staring at the ground uneasily. "You should go into the mountains, *nono*. Slowly. Then you will learn." For *meme* Tashi, engaging with the mountains had been the first step in the path that led him to become a better person. And in his view, the absence of such contact has implications for the state of these mountains today. Turning to me, the old man continues: "The problem today is that people no longer go in the mountains. In those days, people would go in the mountains, and this is why the glaciers were huge."

◆ ◆ ◆

A week has passed since we met with Dolma and Puntsok in Hemis Shuk-pachan. The morning of our climb to Shali Kangri, *aba* Puntsok arrives at our host family's home early, leading the horses. One of them is a colt just a few months old; this will be his first training in high mountain trekking. A mountain guide, local tradition holds, always treks on foot, even when accompanied by horses and riders. *Aba* Puntsok has swapped his woolen jacket for lighter clothes: an off-white shirt, blue pants, and a cap. In the light of the morning sun, he appears older than I first thought he was, moving slowly while he saddles the horses. We are waiting for Stanzin, the student we met the previous week. He had said he'd like to join us, so he could see the glacier of his village for the first time. But instead of Stanzin, his father, Norbu, arrives. A charismatic man in his early fifties, Norbu spent his childhood in the mountains. With a pierced ear, aviator sunglasses, and a white cowboy hat unlike any I have ever seen in Ladakh, he reminds me of his son. It appears that Stanzin's mother did not like the idea of her son going to the glacier, as she is worried that his "body is not strong enough" for the climb. Unlike his father, he has never been herding in the mountains.

I am not especially surprised by this turn of events. As I reported above, Dolma's lament over the lack of knowledge among youths about high pastures and glaciers is a recurring narrative among elders in Ladakh. This perspective is constantly reinforced by comments on young people's physical inaptitude to cope with the agrarian lifestyle of rural Ladakh. People explain that young people have not developed the necessary skills, since today many Ladakhi children are sent to distant boarding schools, while college and university students spend years in cities of India, often never returning to settle in their village. This sometimes spares youths from working the land while visiting their family during summer breaks, and it also prevents them from engaging with the mountains, an activity sometimes described as risky for those unfamiliar with the place.

But the physical inaptitude evoked to describe the younger generation often equally implies, less explicitly, a sense that the character of younger generations is not developed to suit certain aspects of village life. Or rather, as Norbu says to explain his son's absence this morning, "Stanzin is not yet wise enough to do this type of trek." Many believe that agro-pastoralist work, which involves herding in the mountains or preparing fields in spring, requires patience and respect for the mountains, the land, and the animals, and must be performed with a good heart. These attitudes are cultivated over time through experience, and they ensure the success of the activities. As *meme* Tashi would suggest, what is required is the development of a moral selfhood through engagement with the environment.

6.1   On the way to Shali Kangri

After packing lunch in our backpacks, Puntsok, Norbu, and I hit the trail. The mountains of Ladakh are silent to begin with, but the higher we climb, the more the silence envelops us. We pass near an area where a small marshland appears to be struggling to maintain its boggy state, much of the thirsty grass having yellowed. "This *spang* used to be much bigger," exclaims Norbu. "Now look at what it has become!" In Sham, whether it is marshland, spring water, or glaciers, the drying of water sources has become a ubiquitous phenomenon.

As soon as we are beyond the village limits, the landscape changes markedly. The tree line is now behind us, giving way to massive rock formations and tundralike vegetation. The air feels colder. Our group soon passes near an oddly shaped rock, next to which is a vertical prayer flag on a staff. According to a local tale, the rock bears the bodily imprint of Guru Padmasambhava, the founder of Tibetan Buddhism, who is said to have meditated here when he passed through Ladakh on his way to Tibet.[13] Soon, the settlement disappears from sight. Further along, Norbu directs our attention to a land formation on the west side of the valley, a long, narrow, flat-topped hill that overlooks the village. The landscape feature is known as Kong Ka Tal Kan, literally "ridge made flat." On the highest point of the Kong Ka Tal

6.2   Ancient chorten built on the Kong Ka Tal Kan of Hemis Shukpachan village

Kan stands a chorten made of mud and stones, modest compared to many of the chorten built today (figure 6.2). It is said that generations ago there was a disastrous flood, leaving behind massive boulders. The force of the flood raised the soil to create a sharp demarcation in the landscape, or the Kong Ka Tal Kan. It takes me a moment to distinguish this formation from the surrounding hills, but once I spot it, I fully appreciate its immensity, the marks left by the disaster now embedded in the landscape.

Floods are considered by Ladakhis to be a form of retribution, usually sent by chthonic spirits upset by a lack of human morality. After the flood, the story continues, the villagers sought advice from high religious officials in Tibet, who recommended the building of a chorten on the Kong Ka Tal Kan to prevent further disasters.[14] Puntsok remarks, "Our ancestors were really wise, as they listened to the advice from knowledgeable people in Tibet." The ancient chorten that stands on top of the Kong Ka Tal Kan has become quite dilapidated. Talking about it a few days later, two men remark that it is a pity to leave the shrine decaying like that. They had attempted to restore the monument, they said, but nobody helped; they ended their lament with the common refrain: "Nowadays, the children don't do anything."

On our way up, we pass a succession of abandoned farmlands and deserted stone shelters, where, until a decade ago, people would stay during the summer months to process milk products. Although not all summer settlements in Ladakh have become deserted, the one in Hemis Shukpachan, like the one in Likir, nonetheless stands testament to a disappearing way of life in the region. Recollecting her days in this summer settlement as a young unmarried woman, Dolma enthusiastically shared, a few days earlier, how she would spend evenings there with friends and relatives, dancing, playing roles, and sharing stories. This was, she confided, the most pleasant period in her life, when she could spend her days free from the tensions of the village and the household. She had to say goodbye to the high mountains when she married *aba* Puntsok, in order to attend to household matters and raise their children. She truly misses the place: "Whenever I look at the mountains, I think about all this and I dream about those days," she admitted. At the summer settlement, there were both young and old people. "We would provide them [the old people] with drinking water, shrubs and other plants for fuel, and anything else needed, so that they didn't have to do it themselves." Nowadays, with nobody to help them with these chores—because "the younger generations do not care about elders"—old people have no choice but to stop spending the summer in these settlements.

Dolma's account illustrates how the type of knowledge that develops through herding in the mountains—which, in its deeply localized and social nature, would be called by some "traditional ecological knowledge"—is, to paraphrase Paul Nadasdy (2003:63) and his interlocutor in his study of Kluane hunting, "not really 'knowledge' at all," but rather "more a way of life." For the Kluane people, Nadasdy explains, hunting is first and foremost a matter of sociality, an activity that could not take place without sharing, patience, experience, and respect for both animals and humans. These traits form the identity of the hunter and allow hunting to exist. In the same way, knowledge of the high pastures rests on respect for people and animals, and it is equally contingent in nature, emerging from the sociality that informs pastoralist activities.

As we steadily climb, my head begins to spin. I experienced these effects only once before, when I was first in Ladakh. I realize that I am not the only one struggling: the colt is lagging behind, and I can hear *aba* Puntsok trying to catch his breath. As for Norbu, he is taking much pleasure in his role as guide, telling stories linked to the features of the landscape. We arrive at a rocky outcrop on the eastern side of the valley. "This is Mi K'at Sa" (literally, "the place where a man was caught"), Norbu informs us (figure 6.3). As the

6.3   Rock formation in Hemis Shukpachan village named Mi K'at Sa

story goes, long ago a hunter passing this spot saw a blue sheep standing on the rock formation. He climbed to the top by stepping on lichen that had formed natural steps in the rocks. But as the sheep ran down to escape, its hooves broke away the lichen, trapping the hunter on top. To give him the courage to jump down, villagers gathered around, dancing and drinking *chang*. The man eventually jumped off the rock, and died as a result. The story, it is said, serves as a reminder of the Buddhist prescription against harm to animals. A failure to respect this can bring retribution, as in the case of this hunter.

People's profound need to make sense of place leads them to invest landscapes with significance, thus tying the elements that constitute them to people's identity both metaphorically and metonymically (Basso 1996; Dollfus and Labbal 2009a; Thornton 2008). This bond also has practical implications: in instilling self-reflexivity and nurturing an attachment to features of the natural environment, moral conceptions of the land can contribute to shaping ethical dispositions (Basso 1996:40–41; Sivaramakrishnan 2015:1274). In the same way, engaging with the high pastures of Ladakh means delving into the sacred and encountering morality in a material form. The mountains of Ladakh are a repository of moral stories, which serve as a reminder of the disastrous consequences that may follow moral transgressions. For instance, the marks left in the landscape by a flood serve as a reminder of the potential impacts of a corrupted moral attitude, and narratives like the one behind the name Mi K'at Sa suggest the retribution that arises from the transgression of Buddhist ethics for animals. Herders, in narrating landscapes, have long sustained the oral tradition that keeps these forms of knowledge alive.

We reach a place from where we can see a line of magnificent snow-topped summits reaching into an azure sky. Stopping so our horses can graze, we unpack some of our snacks: dry apricots, a few biscuits, and some boiled eggs. Puntsok pulls a bottle of *chang* from his backpack, indulging in a generous sip of the alcoholic beverage. Upon seeing our amused look at his refreshment, the man extolls the virtues of the barley beer to cope with the altitude. Out of breath, Puntsok is visibly affected by the ascent. "I must have aged," he says, smiling shyly, "because the *chang* feels really good today." We all burst into laughter.

As a form of knowledge, intimacy is relational, nurtured by sociality. As Dyson (2015) demonstrates in her study of young herders in the lower Indian Himalayas, the social and cultural practices that characterize movements in the mountains are also fundamental to knowledge acquisition processes, key components of which are friendship and enjoyment. In the same way, accounts of the physical challenges of bringing the animals to the high pastures are often offset by herders' memories of a "very beautiful" and "happy" time. For many, the herding season is associated with a strong sense of liberty, a chance to experience a unique mode of life and to be free from mundane village matters. When speaking of their days in the pastureland, herders often described the pleasure they derived from what is, in its own right, a lifestyle. Time in the mountains is thus not only about labor; it is also about singing, socializing, sharing food and work. The pastures, as herders confided, are also a place where romance blossoms. The intimacy of the high pastures is therefore as much about knowing a place and its resources, the tracks and paths of the mountains, as it is about sighting an ibex or telling jokes while resting in the warm afternoon sun or admiring how a sunbeam illuminates a glacier. This form of situated knowledge is constantly unfolding in and arising from work and from social practices.

While the narratives of the high pasture contribute to its charge, its very materiality also makes it a site of alterity. The sensorial experience associated with an encounter with an ethereal landscape, low oxygen, and extreme physical exertion may translate into a positive or negative experience. As we sit on the ground nibbling our snack, the others enjoy their food, but I have no appetite. My head is spinning like a top. I finally understand the expression "thin air": it feels like every breath is being thwarted by some unknown force. I admit I am upset by this. I have been trekking these mountains for months and now, and for the first time, altitude sickness has me in its grip. Norbu meanwhile, enjoying an adrenaline surge, is ready to resume our trek. I try to repress my anxiety, reasoning that things will be fine; after all, sitting on a horse requires no effort.

We soon set off again. I can barely stay in the saddle. My lungs desperately seek air, even though I am not exerting myself. Before this trek, I had been told repeatedly by people familiar with the physical demands of the mountains to get attuned to my body: to check my heartbeat, assess any form of an oncoming headache, and remain focused. Climbing toward a destination, I was told, is as much about remaining attentive to one's bodily reactions as it is about physical strength. But I cannot concentrate on anything. As I attempt to read my body's signals, reality begins to give way to a dreamlike state. In the far-off distance, I spot the summit of Shali Kangri. Seeing the summit feels both felicitous and frustrating—it is so near but so far. I feel strangled by the cold, and my lungs hurt. Breathing suddenly feels unnatural. Things are blurred. Aware that I am losing control, I give up. I gesture to Puntsok, pointing toward the village. It takes too much effort to speak. While Norbu continues toward Shali Kangri, I despondently descend back to the village with Puntsok, who is obviously disappointed.

DISEMBODIED LANDSCAPES

The experience of the high mountains amalgamates physical strain, bodily sensations, and emotions. This is something herders are aware of and why they advise those unfamiliar with this environment not to take this type of journey lightly. Many elements, both physical and psychological in nature, can explain my predicament while trying to reach Shali Kangri.

Upon reaching my host family, I quickly become the butt of jokes. My host mother teases me, "Shali Kangri, Shali Kangri, you have been talking about this glacier for so many days and now you are already back after such a short trek!" I can only smile half-heartedly in response. Now that I have reached a lower elevation, I feel much better physically, but I am still quite distressed by my failure. Namgyal, who has been waiting patiently but who was also worried, remains silent, well aware of what I am feeling.

I continue to be teased by village acquaintances for the next few days, and feel extremely self-conscious. To the villagers, the reasons for my failure are clear: I was not strong enough for such a quick ascent and should have proceeded more carefully.[15] The incident highlights the fact that navigating high-altitude mountains entails a measure of enskilment, a form of learning through practice, which requires "the fine-tuning of perception and action" and "an education of attention" (Ingold 2000:37, 416). In the mountains, this means not only developing physical endurance, but also being able to read one's physical responses. It is perhaps because this process takes time and practice that elders today see the "part-time" dwellers of Ladakh as having bodies that are unfit to cope with the demands of the place.

But seeing a glacier and experiencing the high mountains is not only a physical journey; it also requires a developed sense of self and of "being in the world." In this way, enskilment in the mountains involves social relations and implies the amalgamation of the person into the relational environment (Pálsson 1994:902). I am well aware that my failure was a product of my stubbornness. Perhaps I was not physically strong enough to climb the mountains, but I also failed by not yet having developed the character required for this enterprise. My incapacity to reach Shali Kangri can be attributed as much to my sense of self as to my body—a reckoning that is not totally foreign to how elders consider the younger generation of today. In fact, it was my plan to do a rapid ascent to Shali Kangri that motivated Norbu to join us that day. Besides his desire to reconnect with a place he had not seen in a long time, I later learned that Norbu felt that going there with an aging man, Puntsok, was not prudent. Aware of what engaging with the mountains entails, he was worried that something might happen.

Whether one meets with success or failure, for many navigating the high mountains is an affectively charged experience. By late afternoon, several hours after I made it back to the village, Norbu returns, smiling, bright-eyed, and content. The journey was both blissful and arduous, he says, especially because the last part of the trek, to the base of the glacier, had to be completed on foot, the terrain being too rough for the horse. The sight of the glacier had a mesmerizing effect and felt like a reward after so much effort, Norbu confides. But there was also something "sad" about the scene, as Norbu puts it, given that the glacier has shrunken significantly since the last time he was in the mountains.

I remain equally captivated by my own journey. Although my experience generated feelings of fear and uneasiness that were the antithesis of Norbu's feelings, both our experiences remain highly visceral ones. What this trek to the high pastures reveals about the knowledge of village glaciers in Sham is that it is not limited to factual observation, but draws from a range of sensations and emotions, developing as it does through a deeply embodied experience.

In late afternoon, we all gather at our host family's house, where a multigenerational group of ten people has gathered to see the pictures of the journey to the high pasture and Shali Kangri. The pleasant aroma of butter tea fills the room. The atmosphere is vibrant, as this will be the first time that most of the attendees will see the glacier of their village. Our group sits around a low carved table. Stanzin joins us, alluding briefly to his inability to accompany us on the trek. I attempt to console him: "It was perhaps a wise decision."

As we view the pictures on my computer screen, Norbu again takes on the role of guide, making the place speak for the people who never went to the high pastures. He describes the various inscriptions in the landscape and shares the stories of Guru Padmasambhava, of the Kong Ka Tal Kan, and of Mi K'at Sa. Strikingly, except for the older observers, few people know either the place-names or the narratives from the high pastures; the abatement of movement in the mountains has resulted in an increasingly disembodied landscape, a process that withers the vitality of local toponymy and lore. The very fact that people are gathering to look at the pictures of Shali Kangri indicates that seeing the summit has become something of an event. This disconnection from life in the mountains does not escape our host mother, who notes, sadly, the abandonment of the summer settlement. When we get to the pictures of Shali Kangri, those familiar with the high pasture all remark, mournfully but calmly, on how small the ice surface has become. The glacier represents the ruin of an earlier, more prosperous era for the high mountains.

◆　◆　◆

Landscapes, in the words of archeologist Barbara Bender (2002:103), "never stand still" but are "time materializing." Yet the temporality of landscapes has often been conceived within the narrow framework of their archival character, in their role as repositories of memories and past events. If, however, landscapes are bound to humans through engagement, the experience of the high pastures of Ladakh tells us that time may also have implications for the nurturing of feelings and sentiments that bond bodies to landscapes. This is a sort of intimacy that the photographic images of Shali Kangri may fail to produce, as is made clear to me when I see *meme* Tashi the following day.

"Did you think the glacier would allow you to see its face so easily?" he asks, bursting into laughter. When I find him, he is sitting in his daily spot next to his house, under the warm Himalayan sun, spinning his prayer wheel mechanically, a blanket on his lap, a thermos of tea at his side. "I told you that you would get *lapotse*," he remarks. "But you should have slept in the mountains. You have been learning for long, by now you should know that it is better to listen to the advice of old people like me!" he exclaims, again bursting into a long bout of laughter. I issue an ashamed "I know," the only answer that comes to my lips. As I pull out the laptop to show him the images of Shali Kangri, his son, a retired schoolteacher, joins us. He explains that he has only seen the glacier twice, because grazing the animals has never been his duty. *Meme* Tashi is pensive as we view the photos. "Has the glacier changed, *meme* Tashi?" I ask. He looks at the screen for some time, pondering. In lieu of an answer, he shares a memory: "Once I reached very close to the glacier in summer. It was hot. The glacier had an ice cave at its bottom. There were many blue sheep, and they were in that cave. The water was dripping on their backs. They must have liked it, because it was really hot. I went inside the cave too, just to see. The water was dripping heavily like a leaking roof. This glacier used to be so huge you would not believe."

Pointing at the image of the glacier on the screen, he remarks that the cave has vanished. "Today these mountains have become dry and empty," he laments. "The blue sheep have no place to cool off anymore." What was once an oasis has now turned into an unfamiliar jumble of ice and bare rocks. The image feels quasi-surreal. "It is as if this was not the same place," murmurs *meme* Tashi. The old man is obviously sad that his advanced age kept him from coming to the mountains with us. "If I could, I would have brought you there and I would have shown you all the paths in the mountains" he says wistfully, adding: "Nowadays, nobody goes to the mountains. Nobody knows this glacier. It is only because I am old that I don't go to the mountains anymore with the animals." I ask *meme* Tashi if he is worried about the glacier vanishing, but his son quickly interjects, explaining that it is not a concern for the moment, since the glacier is still substantial in size—a view I also heard other villagers express. *Meme* Tashi says nothing, something typical of his generation when their children silence them through their perceived sense of superiority, given that, as I often heard, many elders never went to school. This silence, however, speaks volumes about the contrasting sense of obligation the two feel toward the glacier.

CONCLUSION: INTIMATE GLACIERS AND MORAL SELFHOOD

How can interaction with the environment be a transformative experience, one that leads to attachment to glaciers? This chapter opened with Dolma lamenting the younger generation's lack of knowledge about the high pastures and what she perceives to be their lack of concern regarding the recession of the glaciers. The correlation between knowledge about glaciers and a sense of care is regularly alluded to by Ladakhi elders. While these comments fail to recognize prevailing concerns among younger generations, or the existence of different forms of knowledge about glaciers, they nonetheless take on full meaning when we look at how herders of Sham come to know glaciers through their work. For if there is something that distinguishes generational knowledge about glaciers, it is precisely how glaciers are known and which ones are known. The glaciers with which the herders of Sham have become familiar are those that are central to the supply of water in their village. As we saw in chapter 4 and in the recollections of elders in this chapter, they are glaciers with which a community typically feels a strong bond. These glaciers are also entangled in the realm of belief—often imbued with morality—that characterizes the high pastures and that herders encounter through their work, whether as landscape narratives or folk stories.

The knowledge about village glaciers that is produced through herding work is relational, hinging on the sociality that takes place in the high pastures during the summer months. It is also embodied, arising from physical engagement with the mountains and the sensorial involvement it entails. To shed some light on the sensations that may infuse this process of embodiment, the recollection of a trek to Shali Kangri emphasizes how the high pasture can be experienced as a site of alterity. This alterity is constituted through a range of responses, whether to the high degree of physical exertion, a sense of vulnerability, the dizzy feeling created by the high altitude, or simply the feeling of being away from the mundane. Whether this alterity is fraught with anxiety or brings feelings of elation, these responses contribute to the charge of engagement with glaciers. Going to the high pasture has long been so central to people's identity in the villages of Sham that in the absence of a sustained experience of the mountains and of a personal development of knowledge about this place, many elders consider younger generations of Ladakhis no longer fit for this environment. Becoming a moral self in relation to glaciers, in this view, is an act enmeshed in a network of beliefs and practices through which glaciers become intimate, a way of knowing that transgresses "the boundary long presumed to exist between passion and reason" (Rutherford 2016:287).

For herders, work in the high pastures is generative not only of knowledge and affective experiences, but also of ethical reflections. Accordingly, intimate glaciers and their recession cannot be described in generic terms or in an exogenous terminology devoid of emotion. Ultimately, such description may be interpreted, as is done by many elders, as a lack of empathy, since knowing glaciers is for them as much about knowing their physical appearance as about understanding the spheres of morality and reciprocity with which they are enmeshed. The possibility of an ethical relationship between humans and nonhumans may rest on an interface that is imbued with affect and asks for a response (Parreñas 2012:675). Such response may take the form of concern, as expressed by *meme* Tashi when viewing the pictures of Shali Kangri, or of ritual intervention, such as the one initiated by Tsewang Lhamo when she became aware of the recession of Chilchil Kangri. In this way, the experience of the high pastures is constitutive not only of intimate glaciers, but also of a person. To be sure, many people in Ladakh are concerned about the fate of their glaciers, a concern that may be anchored in different modes of being and ways of knowing. But as many elders believe, forms of moral selfhood as related to glaciers are shaped through an engagement with the mountains. This engagement hinges not only on the very act of trekking through the mountains, but also on a sense of being *in* the world, and of cultivating empathy and wisdom, whether through listening to narrative or advice. And what this suggests is that glaciers, as elements of the natural landscape, in emerging from relations that comprise morality, affect, and beliefs, produce enduring subjectivities.

Conclusion

As Glaciers Vanish

To live in harmony with all
Is the essence of morality

—TIBETAN PROVERB

Clearly, the people of today are less fortunate; otherwise, why would
the glaciers shrink?
In the past, the glaciers were enormous.

—TSEWANG DOLMA, NYE VILLAGE

IT IS THE END OF SEPTEMBER. THE AIR HAS BECOME CHILLY AS THE
winter prepares its arrival. Soon, most lucrative activities will be paralyzed,
bringing plenty of opportunities to socialize for some, feelings of loneliness
for others. The day before my departure from Ladakh, in late afternoon, I
find Namgyal sitting in the yard of his home in Leh. The green leaves of the
poplars that abound in this part of the country have started to sport shades
of bright red, orange, and yellow. Ahead, the Stok Kangri summit is slowly
vanishing amid a sky that is awash in shades of pink. In an hour or so, this
sky will be sprinkled with countless stars, a scene obscured by the pollution
found in the capital city of Delhi, where I am heading. "Where will you spend
the winter?" I ask Namgyal. For the past few weeks, with my departure imminent, the question has been vexing him, as it does every year. He has a few
options, including spending the winter in Delhi with friends or joining others in the south of the country, where he can work in a restaurant while
enjoying warmer temperatures. Since his return to Ladakh about a decade

163

ago, life for Namgyal follows the rhythm of work opportunities, with long periods of idleness between summers spent catering to the needs of the tourist industry.

This year, he now tells me, he thinks he will remain in Leh. "Staying here will be long and boring, but perhaps it is the most reasonable thing to do," he concedes. He lights a cigarette and takes a long and pensive drag, before bursting into an asthmatic cough. "I am thinking of opening a business next spring," he continues, "a restaurant. I know well how to serve tourists."

"That's a good idea, Namgyal. I hope you will succeed this time," I reply, trying to give him the encouragement he will not easily find from those around him. Opening a restaurant is indeed something Namgyal has been considering for years, a project that thus far saw light only as a failed attempt lasting several weeks a few years back. But this time, Namgyal feels ready and confident. "I am getting older now; I need to make some plans," he says. "My mother keeps saying that I am stubborn for not sticking to a job. But except for work for the government or in the army, there are no jobs that will keep you busy all year long." Having failed to obtain his high school degree, Namgyal is well aware of the limited options available to him. This failure is also what would have prevented him from entering the army. As one old man once remarked with incredulity, "They could not get enough of us before, and now they are refusing admission."

Namgyal is staring at the horizon. I have a strong sense that his mind is busy reflecting on his own life, which is not an easy exercise, given the questions it brings about the future. "What is tormenting you?" I ask.

"I have been thinking a lot for the past months," he says, confirming my impression. "I need something." He hesitates for a moment. "Even if my restaurant is not a total success, it would be more respectable to do this than failing to commit to something." The experiences of the past year have awakened in Namgyal a desire for righteousness, which for him means achieving a certain work stability and committing to this project, an ideal crafted around what is available to him in terms of skills and knowledge.

◆ ◆ ◆

The following morning, in the early light, I board a plane that, in less than two hours, will bring me from 3,500 meters altitude to near sea level. From my window seat, I can see the glaciers of the Great Himalayan Range, their moraines imprinted in the landscape. High above, the plane follows the enormous, magnificent ice fields in their southward flow toward the Indo-

Gangetic Plain, where the water they release is an invaluable source of life. As I pull my camera from my bag to immortalize the scene, a flight attendant approaches with a warning: "This is a sensitive zone; you are not allowed to take pictures." Her intervention captures the essence of the production of the state in this region since the independence of India: if the fact of militarization is not necessarily apparent in every village, it nevertheless applies to the region as a whole, as these mountains have become a fortress for India. With conflicts between India and both China and Pakistan remaining unresolved, the presence of the army in this border region has become both a primary safeguard of the stability of everyday life and a motor of economic development in Ladakh. If the borderland reality and the fact of militarization in Ladakh are not always experienced as a direct encounter with matters of warfare, the decentering of the agro-pastoralist economy has implications for the entire region, with profound impacts on the fabric of households and community dynamics.

The presence of the state in Ladakh since Independence has changed systematically, from governance from afar, with administrative authorities maintaining only a minimal physical presence in the territory, to a pervasive military presence. In the decades since British rule, the Indian state's territorialization strategy in Ladakh has been oriented toward taming the landscape to safeguard the integrity of its Himalayan frontier, long considered a natural border. The geopolitical context of postcolonial India has legitimatized the military production of the state while consolidating the presence of the state in the region.

Today, Ladakhis live with the inescapable political reality of the border area. This geopolitical situation affects a growing number of nations in South Asia. Amid looming struggles over territorial sovereignty, for many nations of South Asia the exertion of overt control over their borders, which range from impenetrable to porous, plays a significant role in their territorialization strategies (Gellner 2013; Ghosh 2011). If the current borderland scenario in Ladakh is not exceptional, the participation of its inhabitants in the military production of the state may, however, constitute a unique feature of borderland territoriality. It also complicates the often-depicted scenario, following the work of James Scott (2009), of highland areas as zones of state avoidance, revealing how state formation projects and citizens' degree of participation may vary across different topographical landscapes and borderland contexts. As Ladakhis are imagined as scouts of the mountains, they have joined various branches of the state's defense apparatus, opening up new life opportunities through paid work.

But the processes that transform Ladakh today are not associated solely with the production of the state. For the past decades, Ladakhis have also observed changes in their surrounding environment. The recession of glaciers, the depletion of water sources, and the disappearance of vegetation in the mountains are some of the many environmental changes observed today, processes that can be linked to global environmental change. As these changes take place concomitantly with a state production anchored in a borderland logic and a decentering of the traditional agropastoralist economy and way of life, for a generation of Ladakhis the changing environment is interpreted as being intertwined with a changing moral order.

ETHICS OF CARE: MORALITY AND THE MUNDANE

Morality and ethical work do not pertain solely to religious practices and environments. As Anand Pandian (2009) eloquently demonstrates, for men and women of rural Tamil Nadu in South India, morality can be cultivated through mundane activities like farming. Moreover, doctrine does not constitute the sole source of moral orientation for these farmers, who tune their moral compass to various sources, ranging from cautionary tales to popular proverbs, jokes, folk verses, and other popular discourse. This extensive repertoire draws from virtues that date back many centuries and that are contained as much in religious and philosophical texts as in enduring genres of oral verses or in everyday life practices of moral engagement. No matter what contemporary form it may take, this repertoire, in weaving a moral tapestry, makes possible a life of virtue by cultivating ethical inclinations toward ideas of the good (Pandian 2009:15).

Farmers and herders of Ladakh, too, have long cultivated virtue through their work, developing an ethics of care for the land, the glaciers, and the animals. This ethics is rooted in ideas about the interdependence of people, animals, and the elements of the natural environment, and finds material expression in folkloric traditions, storytelling, and ritual practices. But this ethics also develops and is sustained through interaction with the environment, which is generative not only of skills and abilities, but equally of affective experiences. The embodied experience of entanglement within relations of care and reciprocity with animals, glaciers, and the land contributes to the nurturing and sustaining of an ethics of care. Moreover, farming the land and ensuring the continuity of pastoralist activities are seen not only as a moral responsibility, but also as constitutive of persons: many Ladakhis believe that partaking in agro-pastoralist work makes people into moral and caring individuals. If Ladakhis have become immoral, it is not just because they fail in their responsibility toward the land, the glaciers, and the animals, but because they fail to develop a sense of virtue through agro-pastoralist work in the first place. For a generation of Ladakhis, to be a caring individual cannot be dissociated from working the land and attending to the various relationships within the realm of nonhumans that contribute to ensuring prosperity.

The elements that contribute to shaping an ethics of care are, however, not immune to structural and social change. Rituals may fall into disuse as a result of competing temporalities and moral narratives that are imbricated into landscape features and may slowly erode as their transmission is interrupted. The practices that entangle people in relations of care and reciprocity with domesticated animals can also be impeded by a changing economy. The production of the state in Ladakh since the independence of India is such that work in the mountains is increasingly predicated on military activities. Today, many Ladakhis are employed in the army, either as soldiers or in various support positions. Along with the continuing expansion of the state bureaucratic apparatus, the militarization of the region has contributed to pushing agrarian and pastoralist activities to the margins. The organization of agro-pastoralist work is profoundly social and entwined with community and family arrangements. But the fragmented nature of village life today compromises the continuity of community arrangements such as the ones required to sustain farming and herding activities. The struggle to maintain these activities is thus increasingly hobbled by ongoing economic and demographic changes, leaving those who strive to maintain that way of life facing increasingly daunting odds.

But the postcolonial geopolitical context of Ladakh and the work that Ladakhis perform in the production of the state are equally infused with their own values. To be sure, many Ladakhis enroll in the army because they need employment. Ladakhis are also well aware of how dependent they are on the economic spinoffs of India's geopolitical context; given the unresolved tensions with both China and Pakistan, the presence of the army in the region is often seen as helping to safeguard a certain regularity in everyday life. But the military production remains by and large unquestioned, something that must be understood in light of the prevailing idea of collective sacrifice and the fact that the army is imbricated in the intimacy of the family. This acceptance also finds purchase in a military discourse that nurtures affective military labor and in a sense of moral obligation toward the land. Many are quick to point out that the geopolitical context of Ladakh is outside their control but that they have a moral responsibility to protect their land.

For an aging generation, ideals about life wind around agro-pastoralist activities, which are seen in part as a form of a moral pursuit: not only is caring for the land and for domesticated animals a moral duty, but partaking in these activities is seen as allowing people to cultivate ethical dispositions toward others, whether humans or nonhumans, and as creating, more generally, virtuous people. These life ideals are closely associated with rural space and activities. In Ladakh, the major social and economic changes that have been taking place since the independence of India have brought new aspirations. While traditionally, the subsistence-based economy and the organization of work were intricately linked to patterns of household structure, economic changes have brought employment opportunities outside this context. Current individual aspirations among a generation of Ladakhis are, however, tied to the ongoing rural exodus being driven by the restructured economy. The employment opportunities provided by the military and the state's bureaucratic apparatus, along with the education necessary to secure a position in their ranks, are all forcing people to leave the villages. Nowadays, for many, the definition of a successful and happy life is increasingly dependent on these external opportunities.

Crafting one's existence outside the traditional economy also comes with its own challenges and dilemmas. Ladakhi elders often comment that access to paid work and food subsidies along with the abolition of forced labor (*begar* work) has made life easier for younger generations. But for those who choose to build their subsistence around paid work, fulfilling family responsibilities and community obligations in the villages does not necessarily accord with the employment opportunities available, and is therefore fraught with ethical dilemmas. The incapacity to meet expectations for the continuation of the family farm's activities and the inability to provide proper care for aging parents leave many with a bad conscience. Moreover, the decision to make a living outside agropastoralist activities is not necessarily done out of self-interest, as is sometimes perceived, but because it often constitutes the only realistic way to support a family today.

But what the changing economy of Ladakh translates into in the everyday life of elders is a profound sense of loneliness, and that cannot be stressed enough. In our conversations, my elderly interlocutors repeatedly pointed to their deep feelings of isolation, with comments like "There are only old people in the houses today" or "Old people are left alone in the villages." For many elders of Sham, the new demographic dynamics portend the end of continuity. Admittedly, their depictions of desolate villages are not to be taken at face value; rather, they are the product of their social and emotional vertigo. Although many people today spend a great deal of time studying or working outside the rural areas, village communities are still made up of men and women of all ages—farmers, retired army officers, college and university students visiting home, workers on leave, and so on. But in altering the fabric of communal life and in creating new household dynamics, this new demographic reality disrupts the arrangements through which people find support and a sense of purpose as they reach old age. As a result, some feel they have simply become a burden to other family members.

Many Ladakhi elders are confronted by a number of ethical dilemmas. Those who continue to maintain their household's agro-pastoralist activities struggle against growing practical difficulties, including a lack of working hands at home and the progressive dissolution of community arrangements, some of which used to ease the burden of individual work. These new realities result in the growing isolation of elders in the villages, who struggle increasingly with the demands of everyday life matters, sometimes without others to help them. The condition of Ladakhi elders may reflect broader, regional realities: the socioeconomic and environmental drivers of change at play in the Himalayas generally, not just on Indian territory but in other states as well, are bringing about similar changes for many communities where mountain depopulation and demographic change are a reality (Childs et al. 2014).

These issues cannot be dissociated from Ladakhi elders' sense of self. For Tibetan Buddhists, old age is a time to accumulate merit through virtuous acts in preparation for future reincarnation. Yet without the support of a new generation of farmers and herders, many are forced to stop cultivating some of their land or to sell their livestock. When these elders discuss such options, they speak as if they are abandoning a friend. Although many simply have no other choice, their sense of morality is heavily laden with remorse.

AS GLACIERS VANISH

In 1874, J. B. Lyall, settlement officer for the district of Kangra in the Western Himalayas and financial commissioner of the Punjab for the British Raj, described a document relating a royal grant of tax-free land to a family of hereditary astrologers in Lahaul, during the time of Ladakhi kings. Fascinated by the wording of the document, Lyall included the following in a footnote: "The phraseology of the deed of grant was curious. It is drawn up and attested by officials with high sounding titles, signed and sealed at our place, &c., and promises that the grant shall endure till the feathers of the raven turn white and the snow on the mountains black" (Lyall 1874:186).

Especially intriguing to Lyall was the fact that high-ranking, rational men very much like himself had used local folk idioms in the wording of an official government document. In Ladakh, glaciers have long been markers of time, local people being aware that the ice summits are sensitive to human behavior and that their vanishing would be a bad omen for humanity. The allusion to white raven feathers and black mountains is a formula confirming that the land in question was granted in perpetuity. The expression, still used by some Ladakhi elders today, reflects a fundamental reality: snow and glacial meltwater are the source of life in Ladakh. If the mountains were to turn black, the land would have no hope of sustaining life. Glaciers, in the local view, are repositories of time; a vanishing glacier is a trope for the changing condition of humanity.

The practice of nurturing glaciers has fallen into disuse today. Ladakhis' relationship with glaciers exemplifies well the coexistence of the secular and the sacred in an ethics of care. For example, insulating glaciers with charcoal, as related by *ani* Dolma and other elderly Ladakhis during the course of my fieldwork, was a technical intervention and was not, according to my interlocutors, part of any religious ritual. However, the performance of technical interventions need not contradict the belief that a glacier is inhabited by a guardian deity who requires propitiation to ensure that enough water flows in the streams during the sowing season. Thus glaciers are for some Ladakhis simultaneously elements of nature and profoundly social, at once moral and never too distant from the sacred.

When Ladakhis reflect on environmental degradation, they are well aware of the natural causes behind the changes—linking depleted water sources, for example, to snowfall reduction and glacier recession. But such explanations are for them limited, for they do not get at the root of the problem. Assessing change in glaciers is not only about physical processes; it also involves looking at broader changes in humankind, since glaciers are considered a reflection of human morality. If glaciers are receding, many Ladakhis believe, it is because Ladakhis are guilty of wrongdoing both as a society and as individuals.

Ladakhis are not unique in considering glaciers as elements of socionature. Julie Cruikshank provides a fascinating account of colonial encounters in the Mount Saint Elias ranges of North America during the late phase of the Little Ice Age, in which, for the Tlingit people, changing glaciers

C.1   Prayer flags, Tia village

"provided imaginative grist for comprehending and interpreting *shifting* social circumstances" (2005:11–12; emphasis in the original). For a generation of Ladakhis, receding glaciers encompass the materiality of changing moral circumstances, epitomized by a failing ethics of care. This moral landscape started to change with Indian independence and the subsequent war with Pakistan. Many Shammas who experienced the first Indo-Pakistani war believe that this tumultuous historical moment continues to shape Ladakh as a region today and that receding glaciers relate to the still-lingering wounds of that violence. As Robert Desjarlais (2003:25–26) puts it, for Tibetan Buddhists, hardships reflect personal deeds; what people suffer is part of their karma. For many of the elders of Sham, the first Indo-Pakistani war provides the material they need to make sense of the changing glaciers of Ladakh today.

Others believe that the glaciers are receding because people are not taking care of them. Today, glaciers in Ladakh are less visible, whether physically in nature or more figuratively, in the local cultural world. They are becoming anonymous elements of the landscape, their lot paralleling that of Ladakhi elders, who amid the rapid changes that Ladakh has seen since Independence find themselves increasingly isolated in their villages—and as they often remark, nobody seems to care. I believe that I remain faithful to the words of my interlocutors in suggesting that when Ladakhi elders speak of the present time, the time when glaciers recede deep inside the valleys, where fewer and fewer people see them, they are also indirectly saying something about themselves. I believe that when Ladakhi elders talk about the fate of the glaciers of Ladakh, they are also reflecting on their own fate as their presence and influence decrease amid the dazzle of a new era.

GLOSSARY OF LADAKHI TERMS

Ladakhi terms are spelled as pronounced in Sham. Where known, transliterations of the Tibetan spelling following the Wylie system are in parentheses.

aba   (a ba)   father
abi   (i pi)   grandmother
ama   (a ma)   mother
amchi   (am chi)   practitioner of Tibetan medicine
ani   (a ne)   aunt
azhang   (a zhang)   uncle

chang   (chang)   local beer, made from fermented barley
chorten   (mchord rten)   stupa (Buddhist monument)
chumik   (chu mig)   spring water

doksa   ('brog sa)   high pastures
dzo   (mdzo)   cow-yak hybrid
dzomo   (mdzo mo)   cow-yak hybrid female

geshe   (dge bshes)   the highest degree in the Gelug school of Buddhism;
in common usage, also the name given to a monk who has obtained such a degree
goba   ('go pa)   head of the village

jhonsa   (bzhos sa)   place in the pasture where dairy activities take place

kangri   (gangs ri)   glacier
Karja   (Gar zha)   Lahaul, a region of North India that today falls under the administration of Himachal Pradesh state and the district of Lahaul and Spiti
khangpa   (khang pa)   the ancestral house

khangu   (khang chung)   subsidiary house

kholak   roasted barley flour mixed in butter tea

kishang   (khyi spyang)   a hybrid of a dog (*khyi*) and a wolf (*spyang ku*)

komnyer   (dgon gnyer)   caretaker of a monastery

koncha   (gon cha)   traditional long woolen coat (for men)

konjok   (*from* dkon mchog gsum, literally "the three rare and supreme
     ones") an exclamatory expression often employed in Ladakh, referring
     to the triple gem of Buddhism: the Buddha, his teachings (Dharma), and
     the community (Sangha)

labrang   (bla 'brang)   monastic treasury

lama   (bla ma)   spiritual teacher in Tibetan Buddhism

lha   (lha)   spirit, deity

lha tho   (lha mtho)   altar dedicated to local deities

Losar   (lo gsar)   new year

lotho   (le tho)   almanac

lu   (klu)   spirits of the underworld

makpa   (mag pa)   groom, in Sham; sometimes used to refer to a married
     man in a context of uxorilocal residence

mane   (ma ni lag 'khor)   prayer wheel

meme   (mes mes)   grandfather

mo gos   (mo gos)   traditional long woolen coat (for women)

nalbu   (nyal bu)   illegitimate children

nono   (no no)   younger brother

nyingpa   (rnying pa)   old

onpo   (dbon po)   local astrologer, ritual specialist

perak   (be rag)   turquoise-studded headdress

phu   (phu)   high pasture, situated at the top of a valley

pu lu   basic stone shelter

romkhang   (ro khang)   burial place; also a funeral hut

rong   (rong)   gorge

sadak   (sa bdag)   subterranean spirits of the earth

sang   (bsangs)   offerings of juniper incense

shanku   (spyang ku)   wolf

skitpo   (skyid po)   enjoyable, pleasant

skyin jug   (skyin jug)   during a marriage, bride's lament at leaving
     her family

smanmo   (sman mo)   female ghosts

soma   (so ma)   new

spang   (spang)   grassy wetland; meadow

spao   (dpa 'bo)   a hero

storma   (gtor ma)   dough creations used in many Buddhist rituals

tangmo   (grang mo)   cold

thang   (thang)   arid plain

tsampa   (rtsam pa)   roasted barley flour

tsokpo   (btsog po)   bad, dirty, reprehensible

yul lha   (yul lha)   territorial god; sometimes referred to as a village god

zhidak   (gzhi bdag)   literally, "foundation lord": a territorial god, often associated with mountains

NOTES

FOREWORD

1   Ingold 2011; Kohn 2013; Latour 2014; Haraway 2015.
2   Descola 2013; Steffen et al. 2011; Bonneuil and Fressoz 2016.
3   Piketty 2014.

PREFACE

1   On similar sentiments expressed by elders in the Himalayas, see Desjar-
     lais (2003).

INTRODUCTION

1   The limits of Sham (from *gsham*, meaning "lower" in Ladakhi)—also
     referred to as Sham Ilaqa ("area" in Urdu) or sometimes "lower Ladakh"
     in the literature, or even "Sham Valley" by Ladakhis—appear to be fluid
     (see, e.g., Dollfus 1989:22–23; Rizvi 1996:2). While most Ladakhis suggest
     that the area extends from Nyemo to Khaltse, I have at times encoun-
     tered the idea among Ladakhis that it extends farther still, as far as the
     village of Achinathang, and perhaps even the border of Pakistan. The ter-
     ritory of Sham covered in this study extends from Nyemo to Khalatse.
     My study focuses on the predominantly Buddhist population of Leh Dis-
     trict, and none of the villages in Sham where I conducted research had
     inhabitants of Muslim confession.
2   Ladakhi terms are spelled as pronounced in Sham. Tibetan spelling
     following the Wylie system of transliteration is provided in parentheses.
3   While details on the history of the practice in Ladakh may be scarce, infor-
     mation on glacier grafting in Baltistan—a neighboring region in today's
     Pakistan with close links to Ladakh dating back centuries, particularly

through trade (Rizvi 1999)—sheds light. The earliest documented reference to glacier growing in Pakistan is attributed to British colonial administrator D. L. R. Lorimer, who observed it in Baltistan in the 1920s (Buddruss 1993:77; see also Israr-ud-Din 1960). Buddruss (1993:79) explains that Shina speakers in the region talk of the cultivation of glaciers as "causing ice to sit," in the sense of "to remain." In his analysis of an undated text written in the Shina language, Buddruss (78–80) also reports the use of charcoal, in this case put into a ditch and layered with gourds of water, ice, and straw, the entire assemblage then buried in the ground and eventually growing ice. This technique was also observed by Ingvar Tveiten (2007), who provides a comprehensive study of glacier grafting in Baltistan, where he conducted research among glacier growers. Buddruss's and Tveiten's findings shed light on what the practice may have involved in Ladakh (see also Faizi 2007). Whether the past practice of glacier cultivation in Sham was similar to those observed in Baltistan is likely to remain unanswered in detail, but it is reasonable to assume so. Mentions of the use of charcoal in both Baltistan and Ladakh strengthen this assumption and, additionally, confirm that the practice served practical, rather than strictly religious, purposes, in contrast to rituals for glaciers in the Himalayas described by others (see, e.g., Norbu and Harrer 1986:48–50).

4   There have been recent initiatives directed at building artificial glaciers in Ladakh to address current issues of water stress. These initiatives have, however, been undertaken by local organizations or individuals. For more on this, see Gagné (2016).

5   The ecological implications of Buddhism have inspired a wealth of writing, with topics ranging from eco-Buddhism, to deep ecology, to Buddhist environmentalism. In foregrounding an ethics of care, my approach departs from this scholarship, which generally takes Buddhist texts as its point of departure and focuses on aspects of Buddhist tradition that inform environmental ethics. However, in this literature, Buddhism too often remains an ideology with little connection to the pragmatic reality of everyday life, as Huber (1997) points out in his critical perspective on "green Buddhism." Moreover, given that textual orientation to Buddhism is but a recent development in the region, the emphasis on scripture provides only a fragmentary window into ethical dispositions toward nonhumans in Ladakh.

6   See Foucault (1976, 1994). This scholarship distances itself from the Kantian ethics of duty, according to which moral actions are performed out of obedience to collectively defined rules and norms, and focuses instead on the Aristotelian ethics of virtue, according to which moral acts are an outcome of reasoning and deliberation. See, for example, Daswani (2013), Dave (2012), Mahmood (2005), and Stafford (2013).

7   The various contending points of view have been amply discussed else-where. See, for example, Laidlaw (2002), Lambek (2008), and Zigon (2007).

8   Local deities and spirits in Tibetan Buddhism have been the object of detailed examinations: see Buffetrille (1998), de Nebesky-Wojkowitz ([1956] 1993), Dollfus (1997), Karmay (1996), Pommaret (2004), Ramble (1997), and Riaboff (1996).

9   Dollfus (1989:131–33) analyzes how the Ladakhi household reflects a cos-mological order, elements related to the sacred being located on the upper floor, the profane on the lower one.

10 On the relation between chthonic beings and the social realm in Ladakh, see Butcher (2013b:165–170), Day (1989), Mills (2003), Phylactou (1989), and Srinivas (1998). Pirie (2007) departs from this view and maintains that for Ladakhis "neither the moral order of Buddhism nor the realm of the spirits is related to the moral and political order" (90). Rather, in a Durkheimian fashion, Pirie postulates that the social order *in itself* is sacred and to be found in "peaceful and harmonious relations between people and their households" rather than in relation with local spirits (130). Pirie insists that for Ladakhis, the moral order "is a human order, one found in the activities of individuals, rather than a cosmological or divine order" (125–26).

11 Ideas about nonhumans, it should be noted, do not emerge only from an orthodox form of Buddhism, but also from "the religion of men" (*mi chos*) or "folk religion," which is distinct from the "religion of the gods" (*lha chos*) (see Stein 1972:165, 191–229). Stein (1972:192) calls this form of worldly religion the "nameless religion," which is "always uttered in a poetic style characterized by the use of metaphors, clichés and proverbial sayings." Both forms of religion are central to how Ladakhis dwell in the world, and as Butcher (2013b:8) aptly points out, in Ladakhi Buddhist practices their boundary is never really distinct.

12 The "affective turn" in social sciences has its roots in the work of social theorists who have drawn from the writings of the seventeenth-century philosopher Baruch Spinoza, particularly the interpretations developed by Gilles Deleuze and Félix Guattari. See Clough (2008), Gregg and Seig-worth (2010), and Massumi (2002).

13 As a number of commentators (e.g., Laszczkowski 2016; Navaro-Yashin 2012; Thrift 2008) have pointed out, this dimension of imagination has frequently eluded analyses based in actor-network theory.

14 For a review in the context of South Asia, see Sivaramakrishnan (2015).

15 By "production of the state" I am referring to the practices and forms of imagination through which the Indian state is built and performed. Modes of production are diverse and their logics can differ from one state to another within a nation-state; the case of Ladakh shows how this production is shaped by geopolitical concerns and by mountainous topographies.

16 See also Gold and Gujar (2002) for an account on the apparently paradoxi-
cal representation, formulated amid rapid social change, of time under
an autocratic rule as happy.

17 Haraway 1989; Orlove 2002; Strathern 1988. See also Campbell (2013) for
a discussion specific to the Himalayas.

18 Kohn 2014; Latour 2013; Tsing 2015.

19 On the compatibility between secular and sacred geography in the
Tibetan Buddhist context, see Salick and Byg (2012) and Samuel (1993).

20 To account for this view, which implies an understanding of the world
and a view of nature that differ from the Western intellectual project,
we may speak of a different ontology; see Descola (2013), Kohn (2013),
and Viveiros de Castro (1998).

21 According to the Census of India of 2011, Leh District has a Buddhist
majority (66 percent) and a Muslim minority (14 percent).

22 With three persons per square kilometer, Leh District has the lowest
population density in India.

23 The Indus and Shayok, the main rivers crossing Ladakh, sustain only
10–15 percent of Ladakh's agriculture (Shaheen et al. 2013:24).

24 In July and August, the mean monthly values vary between 12.55 and 23.7
degrees Celsius, while in January and February average temperatures
range between −15.6 and −5.5 degrees Celsius (Dame and Nüsser 2011).

25 This means that during my main research stay, I could not go beyond
Khalatse village to the west and I could not go to the Nubra Valley or the
Changthang area.

26 Data were extracted from Dawa (1999:373) and Government of Jammu
and Kashmir (2012).

27 A *dzo* is a male cow-yak hybrid. A female cow-yak hybrid is a *dzomo*
(*mdzo mo*).

28 Gellner (2013) notes how these processes are linked with the tracing of
postcolonial borders that followed the creation of the new nations of South
Asia. Thus, although Ladakh was arguably a border area under Dogra
rule, the experience of this border, which was poorly enforced in Ladakh
itself, differed in many ways from the experience of a border area following the tracing of postcolonial borders in South Asia.

CHAPTER 1: THE LONELINESS OF WINTER

1 Lack of income in winter is generally mentioned by those who make a
living in the tourist industry. Tourism, though significant for the local
economy, is limited to a four-month season and does not provide a stable
source of income. Moreover, the service economy associated with tourism has made only limited inroads into Ladakhi villages, where it often
plays at best a trivial role (see also T. Tsering 2008). In contrast, serving

in the army or working in the civil service provides a year-round income and stable remittances for family members.

2   The four seasons of Ladakh are of unequal length. *Spit* (*dpyid*), or spring, is short, lasting from April through May; *yar* (*dbyar*), or summer, begins in June and lasts through September; *ston*, or autumn, encompasses the months of October and November; while *rgoun* (*dgun*), or winter, sets in in December and holds through March (Dollfus 1989:25).

3   In winter, the price of a kilo of any vegetable (if it is even available) can be up to five times its summer price (see Dame and Nüsser 2011:186).

4   The preciousness of firewood in winter is well encapsulated in the Ladakhi saying "The lack of firewood is more serious than the lack of flour" (Khan 1998:172)

5   For many, escaping the region for the winter before the roads close is of great concern. Ladakh remains accessible by plane in winter, but the price of flight tickets has seen a sharp increase in recent years.

6   Hair was once a status marker distinguishing peasantry from nobility, as only members of the royal lineage could wear their hair entirely long, a tradition that many Ladakhis elders still observe.

7   Cunningham [1854] 2005:19. Ladakh effectively lost its independence with the signing of the Treaty of Leh (1842) by Tibet, which acknowledged the Dogra (a Hindu Rajput dynasty founded by Gulab Singh) suzerainty over the region (Aggarwal 2004:33). The history of Ladakh prior to the Dogra conquest is characterized by numerous attempts by various powers to bring the region under their sway (Rizvi 1996, chapters 4 and 5). See Aggarwal (2004:32–36) for a review of the key elements of the political history of Ladakh from the Dogra conquest to the independence of India.

8   Aggarwal and Bhan 2009:521. Fifteen years earlier, in 1819, Kashmir came under the rule of the Sikh emperor Ranjit Singh. During the first Anglo-Sikh war (1845–46) Gulab Singh, who as one of Ranjit Singh's vassal princes had been instrumental in the conquest of Kashmir, sided with the British after their victory and was able to purchase the entire territory of Kashmir and surrounding areas for the relatively small sum of 8.5 million rupees (Gutschow 2004:24). This gave rise to the political unit of Jammu and Kashmir.

9   Subsequently, the British asserted greater control over regional trade, through the Commercial Treaty of 1870. If this placed greater emphasis on surveying the territory, the physical presence in this vast territory of any administrative authorities or of the Dogra army was nominal and often limited to the few comfortable months of the year. Hence, although the British joint commissioner was in Ladakh year-round due to an appointment in Baltistan—which until 1947 was included in the Ladakh *wazarat* (district)—in what is today Leh District this physical presence remained limited.

10	Little is known about the economic conditions under which Ladakhis lived before the Dogra rule, but observations by missionaries and British officials suggest that, although Ladakh was not an egalitarian society, Ladakhis seemed to be more materially comfortable—insofar as material comfort can be said to characterize a peasant society of this time—before than after the Dogra dynasty took control of the region in the 1830s (Sheikh 1999:342).

11	The conditions and modalities of this work have been analyzed extensively (see, e.g., Bray 2008; Grist 1992).

12	A *tehsildar* is an administrative revenue officer responsible for tax collection in a given *tehsil*, or administrative division.

13	The name Khachul, or *Kha che yul*, the "land of the Muslims," is often used by Ladakhi elders for Kashmir.

14	The Nangso were the king's formal treasurers; see van Beek (1996:121), citing Ramsay (1890).

15	The *anna* is a now-defunct monetary unit that equaled one-sixteenth of an Indian rupee.

16	*Abi meme tus* (*i pi mes mes dus*) or "the time of the grandmothers and grandfathers" is an expression often used by Ladakhis to designate something dating to previous generations.

17	But elders' accounts also suggest that peasants found various ways of skirting these abusive practices, for instance by removing weights from the scales or by "bribing" official employees by putting treats such as *kholak* (roasted barley flour mixed in butter tea) on top of their bags of grain.

18	In colonial India, European men and women of authority were addressed by the polite title *sahib*, a reference still used to describe past interactions with people of authority or a certain social status.

19	In Sham, these sundials or sun markers are also referred to as *tegor*.

20	See Khoo (1997) on solar observations in Ladakh.

21	The 1970 flood in Nyemo is said to have been caused by flood after a part of the glacier above the village collapsed and fell into a lake below.

22	The emic perspective here refers to the epistemological perspective of the community who practice polyandry. The question of polyandry in the Himalayas has been the object of several anthropological studies (Berreman 1980; Crook and Crook 1994; Goldstein 1978). Levine (1988) also formulates a cultural argument that challenges functionalist perspectives on polyandry.

23	As the story goes, aware that Ladakhis of Muslim confession were poor in terms of agrarian assets, rulers feared that Buddhists would meet the same lot if they did not take measures to prevent land partitioning. The practice of polyandry came to be seen as a duty that was fundamental to the survival of Buddhist Ladakhi society and, by extension, integral to its culture.

24 Karja corresponds to the region of Lahaul in today's district of Lahaul and Spiti (Himachal Pradesh state). Some Ladakhis had ties in the region, which has a significant Buddhist population.

25 This type of arrangement generally occurs in families with only female descendants, who need a male to ensure the continuity of the household lineage.

26 Although marginal in number, I observed cases of polyandry and polygyny in some villages going back one or two generations (see also Bhan 2014:35). On polyandry in Ladakh, see Berreman (1964), Crook and Crook (1994), Crook and Shakya (1983), Gutschow and Gutschow (1998:117), Hay (1997), and Prince Peter (1963).

27 Polyandry was banned in 1941 with the passing of the Buddhist Polyandrous Marriages Prohibition Act, but it was not abandoned systematically and continued to prevail for some time. The act was followed in 1943 by a law banning primogeniture, the Ladakh Succession to Property Act. See Bertelsen (1997) for a discussion of the history and the rationale behind the ban on polyandry. While disadvantaged younger brothers in some villages welcomed the ban on polyandry and primogeniture, the measure created much resentment among first-born males (Aggarwal 2004:72).

28 In the local lexicon of Sham, polyandry is referred to as *thunte (mthun te tsoks) yarte (g yar te tsoks)*, or "in the way of an agreement, in the way of a borrowing," which also has the contextual signification of "compromise, sharing" or "living together in mutual understanding." In Ladakhi polyandrous unions, the marriage celebration included only the eldest brother, but because his wife was joining the household, his younger brothers were de facto joining the union. The definition of polyandry as *thunte yarte* makes it clear that the elder brother had the upper hand in this arrangement.

29 Moreover, as elders told me, "access to the wife" was not always an easy matter, especially in a union that included many brothers. Hay (1997) provides one of the very few accounts on women's perspective on polyandry in Ladakh.

30 The man is alluding to *tsogspa (tsogs pa)*, village bodies that supervise a number of activities at the community level, for instance the cleaning and repair of water canals.

31 Issues related to the depopulation of the Himalayas and its threat to a family-based care system for the elderly have been discussed for Nepal (Childs et al. 2014; Goldstein and Beall 1982) but remain largely unexplored for Ladakh (see Jordan and Sen 2014 for an exception).

32 According to the Census of India (2011), most Ladakhi farmers (76 percent) possess 1 hectare or less of land.

Proverb: From Lhamo Pemba (1996).

1   *Pathan* is the Hindi word for men of Pashtun origin.

2   Primarily of Tibetan descent, since about the fourteenth century the population of Baltistan, today a region of Pakistan, has been Muslim. The terms Balti and Khache are commonly used by Shammas to distinguish not only between Shia and Sunni Muslims, but also between people from Baltistan (Balti) and people from Kashmir (Khache).

3   The State Forces were troops that had been recruited by the princely states of the British Indian Empire. They operated alongside the Indian army.

4   Among the few sources to discuss the invasion of Ladakh are Chibber (1998), Kaul and Kaul (1992), Prasad and Pal ([1978] 2005), and Sen (1969). Gutschow (2006) provides first-hand accounts of the Partition in Zanskar. First-hand accounts in the Ladakhi language can be found in a special edition of *La dvags kyi Shes rab zom* [Sheeraza Ladakhi] 20, nos. 3–4 (1998–99). There are, however, no written sources on the perspective of the Muslim Ladakhis from Purig area who joined the raiders. Their testimonies would greatly contribute to broadening the scope of the available accounts of the war, which remain by and large centered on the perspective either of Buddhist Ladakhis or of the Indian state.

5   This book does not examine the events of the Partition that took place in other parts of Ladakh. In the Nubra Valley, villagers organized a resistance without any initial support from the Indian army. There, at the young age of seventeen, Chewang Rinchen organized the Nubra Guard, successfully mounting several attacks. For an account from the Nubra Valley, see Verma (1998).

6   Mona Bhan (2008:146) explains that older people among the Brogpa of Ladakh—a community living along the Line of Control in the districts of Leh and Kargil, of both Buddhist and Muslim confession—also remember doing *khral* (mandatory work akin to *begar*, established during the era of the Ladakhi kings) for Pakistan during the first Indo-Pakistani war.

7   A *perak* is a large headdress ornamented with several gemstones (mainly turquoise) traditionally worn by Ladakhi women.

8   *Storma* (*gtor ma*) rituals are generally known as *skurim* (*sku rim*) and are "a frequent and important part of village life, signalling the people's preoccupation with the dangers of evil spirits" (Pirie 2007:98). *Storma* is the name generally given to dough creations used in many Buddhist rituals, which are offerings to remove malign influences (Mills 2003:190–91; Samuel 1993:265–66).

9   Interestingly, none of my interlocutors referred to the work done for the Indian army as *begar* work.

10   Livestock were killed not only for food: when the bridge over the Indus in Khalatse was burned, the raiders killed *dzo* in large numbers in order to make coracle boats with their skins.

11   *Lha tho* (*lha mtho*) are altars dedicated to local deities; their structure often incorporates the horns of goats and ibexes. *Konjok*, an exclamatory expression often employed in Ladakh, refers to the Triple Gem of Buddhism: the Buddha, his teachings, and the community (*kon chug sum*).

12   See also Gutschow (2004:27). This does not mean that Ladakhis of Sham would never kill domesticated animals for consumption. However, they consumed meat moderately or saved it for special occasions.

13   For various hypotheses, see Akbar (1999), Prasad and Pal ([1978] 2005), and Sen (1969).

14   The name Balti rong (literally, the "the narrow valley of the Balti") is sometimes used to refer to Purig area toward Kargil, south of the Indus. The area includes several villages, among them Chiktan, Shagkar, and Mulbekh (Rizvi 1999:109). Occasionally, though, my interlocutors would use "Balti rong" to refer exclusively to Chiktan.

15   The lexicon used by Shammas reveals that the raiders are not seen as a unitary group. In conversation, elders referred to the invaders in various ways: they used either the metonym Pakistan, the name Pakistanpa (people from Pakistan), the term *kabayilis* (which means "tribesmen of the mountains" in Urdu), or the term *phyi pa*, which means "outsiders" in Ladakhi. Yet when referring specifically to the Muslim Ladakhis who joined the raiders, they would use the expression Baltis or Chiktanpas (people of Chiktan).

16   These relations of mutual esteem were perhaps reinforced by the fact that they were carried down through generations, as traders would take over the business of their fathers. I have, however, never seen it mentioned that these trading relations were formalized in ceremonies of friendship, as von Fürer-Haimendorf (1988) observed among Sherpa communities, nor is it reported by Janet Rizvi (1999) in her extensive studies of Ladakhi traders.

17   According to several interlocutors, this insult referred to the belief that Buddhist Ladakhis would eat carrion, something that Muslims strictly refrain from.

18   Janet Rizvi (1999:105–6) also reports that traders coming from Purig area and from Skardu to the annual salt trade in Sakti near Leh were extremely poor. Shammas' idiomatic expressions also allude to this poverty. For instance, elderly Shammas still say: "It is like the Balti got salt!" to refer to someone who is enthusiastic about being offered something. Another saying asks, "What does a Balti not need?" implying that a Balti has nothing at all.

19   Discussing ethnic and ethno-nationalist conflicts, Stanley Tambiah (1997:1177) points out that the sudden open animosity between communities that lived together rather peacefully may emerge precisely because of extant tensions, and this despite apparent cordial interactions. More than ethnic difference, he maintains, at the core of such tension lie contests over status and alleged economic well-being. This seems to have been a repressed element of friction between Buddhist and Muslim Ladakhis who would interact in Sham prior to the war.

20   The war took a communal turn not only in Sham, but in Leh as well. However, it should not be assumed that all Muslim Ladakhis shunned the Indian army and aligned with the Pakistani raiders; some, in fact, joined the resistance (Kaul and Kaul 1992:169–73). Also, given that work for the armed forces during the war was sometimes rationalized as part of an already existing system of forced labor (see Bhan 2008:146) and that villagers were not always aware of the different national identities established with the Partition, we cannot assume that all villagers acted under patriotic feelings and clear notions of allegiance.

21   The fact that Ladakhis rose against the Pakistani invaders remains difficult to reconcile considering that, as we saw in chapter 1, after the initial rebellions against the establishment of the Dogra rule they by and large conformed to the authority of the successive Dogra rulers. But as Jacqueline Fewkes (2009:45–46) explains, religious affiliations were only of incidental interest to the Dogra ruler (see also Rizvi 1999:50). Rather, "a desire to participate in Ladakh's trade in pashmina with Kashmir was the primary reason that Gulab Singh's forces invaded Ladakh." (Fewkes 2009:45). Moreover, Ladakhi history remembers the conquest by Zorawar Singh as an act of plunder, when treasuries and monasteries were looted (Tsering Samphel 1997). But religious conversion in Ladakh was apparently never a dominant concern for the Dogra. As Aggarwal (2004:32) explains, by and large, the successive Dogra rulers' "attitude toward their Ladakhi subjects was extractive, autocratic, and aloof." Kaul and Kaul (1992:107) link the Dogra rulers' disinterested attitude toward Ladakh to its geography, its harsh climate, its difficult access and lack of communication. Ladakh therefore came to be "regarded as an encumbrance, a liability, rather than an asset." Overall, "a smug complacency with regard to this part of the State became the prevailing attitude of the authorities in Srinagar," and the appointment of administrative officers to the region was in fact treated as a punishment for those who had previously proved to be unfit or corrupt (Kaul and Kaul 1998:107).

22   Mona Bhan's (2008) work on the allegiance to the Indian state during the war is also enlightening. Regarding "the occupation of several parts of Ladakh in 1948 by Pakistani soldiers," Bhan cites a Kargili informant

who explains that this "was a time when Kargilis realized that they would be better off with India than they would be with Pakistan," on the basis of their poor organization during the war and the fact that they were extracting resources from the villagers. The interlocutor asserts, "Kargilis decided that even though India was predominantly Hindu, it had discipline, it had democracy. Pakistani occupation of Ladakh in 1948 was taken as a curtain raiser to the kind of life Ladakhis would live in Pakistan." Bhan maintains that this historical episode "was salient in establishing relationships of border communities in Ladakh with the Indian state" (2008:147) Given the emphasis my informants put on the abuse they were subjected to during the war, as an outlook on "the kind of life Ladakhis would live in Pakistan," this no doubt constituted a decisive factor in the mobilization of Shammas alongside the India troops.

23   Given the existing bonds of kinship between Buddhists and Muslims in Ladakh (see, e.g., Fewkes 2009; Smith 2011), it may come as a surprise that Buddhists sided with the Indian army. In fact, Ladakhi scholar Abdul Ghani Sheikh (1998–99) reports that during the conflict, "silent pacts were made between Buddhists and Muslims, pledging to shield each other if either Pakistan or India was victorious" (cited in Aggarwal 2004:37). None of my interlocutors alluded to a similar pact; this could be explained by the nature of the relationship between Buddhists and Muslims in Sham, which differed from the relationships between the two communities in Leh, during the time of the Partition. For instance, intermarriage in the context of trade was common among the Arghun community: Muslim traders from different regions along the Silk Road would marry Buddhist Ladakhi women from Leh (Fewkes 2009). But this type of alliance was not found in Sham, according to my interlocutors. Today, the area of Sham that lies between Nyemo to Khalatse (see chapter 1) is inhabited exclusively by Buddhists. None of the villages where I conducted research had Muslim families at the time of my fieldwork, but before the war, Basgo, Nyemo, and Saspol each had several Muslim households. Based on various accounts, the three villages together comprised no more than ten Muslim families and these families are said to have fled with the raiders when they retreated back to Pakistan at the end of the war. According to my interlocutors, there was no intermarriage between Buddhists and Muslims among these families, and there are also indications that some of these Muslim families were ostracized by the rest of the villagers. While the discourse of my interlocutors on the question may be colored by the current tensions that prevail between Buddhists and Muslims in Ladakh, the situation nonetheless suggests that the nature of intercommunal relationships in Sham was different than in Leh. The idea by which Buddhists are seen as having the prerogative over a place also finds echo in spatial distinctions made by Buddhist

Ladakhis. Ravina Aggarwal explains that the village of Achinathang, which is at the westernmost part of the extended limits of Sham (see chapter 1), at the edge of Leh District, and which also borders the Purig region in Kargil District, is referred to by Buddhist Ladakhis as *phyi nang gi sa mtshams*, meaning "the border between insiders and outsiders." This border also loosely represents a religious divide, the district of Leh having a Buddhist majority and the district of Kargil having a Muslim majority (Aggarwal 2004:14–15).

24 On communal conflicts in Ladakh, see Aggarwal 2004; Bertelsen 1996; Deboos 2013; Smith 2013; Srinivas 1998; van Beek 1996.

25 According to my interlocutors, Muslim Ladakhis who sympathized with the raiders received extremely harsh treatment from the Indian armed forces and from some Buddhist Ladakhis. In some instances, this also extended to Muslim Ladakhis who had not joined the raiders, as described in this chapter. As communal tensions reached a critical stage, influential members of the Buddhist community in Leh urged the Indian army to put an end to this violence. Many interlocutors reported how years after the war, at the latter stage of his life, Hari Chand—a Lahauli officer who made a great impression on Shammas—came to Ladakh and lamented that he had been prevented from "chasing all the Muslims away from Ladakh" during the war, which led to a demographic shift in the region. This constitutes a troubling assertion, given the social and cultural heritage of the region; see Smith (2012) on this alleged shifting ratio. Communalism, in demarcating the two communities along religious lines, is part of the various strategies deployed by political leaders of Ladakh today (Aggarwal 2004; van Beek 2001, 2004).

26 In Ladakh, a *labrang* (*bla 'brang*) is a monastic treasury; here the name is used to refer to a building used in the past to shelter animals and store goods such as butter, wool, and grain collected from villagers for the monastery of Hemis.

27 In 2010, Ladakh was affected by disastrous floods that claimed several hundreds of lives following a cloudburst.

28 On the Muslim families of Sham, see note 23.

29 Although not necessarily framed as a consequence of the first war with Pakistan, the question of the growing chasm between Buddhist and Muslim Ladakhis has been the object of a number of studies. See Gutschow (2006), Smith (2009, 2012, 2013a, 2013b), and van Beek (1996).

30 Here, *meme* Rinchen refers to the Ladakhis who were guarding the mounted gun in the Hemis Labrang. Accounts of the number of people executed differ, though it never exceeds five.

31 In her account of the events of the Partition in the Zanskar, Gutschow (2004:24–28) reports cordial relations between the raiders and the local population: "There were no stories of rape or other horrific abuses which had plunged other border regions into panic and terror during this same period" (27). Testimonies from Sham, however, contradict this observation, possibly because Ladakhis from the Balti rong did not reach Zanskar; according to my informants, it was men from this community who were guilty of rape.

32 Strategies intended to undermine and destabilize the status of the (sometimes sudden) enemy in times of conflict, abduction and rape threaten or prevent cultural continuity by violating family life and invading the inner social space. See Das (2007) and Tambiah (1997:1177–88).

33 *Chos* refers to Buddhist religion; in this context, *meme* Rinchen implies that he was not well aware of the code of ethics of Tibetan Buddhism.

34 A chorten is a Buddhist monument and reliquary.

35 Bhan (2014:1–3) also reports that the Brogpa of Kargil District in Ladakh believe that the recent Kargil war has angered the local deities. See also Butcher (2013b:165) on local perceptions of the link between the independence of India, people's changing attitudes, and recent natural disasters.

36 Many Shammas believe that in dying at the young age of thirty-seven, Khushal Chand suffered retribution for his alleged involvement in the killing of the innocent people near the bridge.

37 The taming of forces that dwell in the landscape through the construction of architectural structures like chorten as a means to stabilize negative geomantic influences has deep roots in Tibetan mythology (see Samuel 1993:168; also Gyatso 1987; Mills 2006, 2007). The chorten near the bridge in Khalatse has a similar function. Butcher (2013b:180) also notes that the construction of a one-hundred-foot-high, gold-plated statue of the Maitreya in the Nubra Valley of Ladakh was meant both to protect the people of the valley from increasing water problems (deluges and shortages) and to ensure peace, since the valley borders on both China and Pakistan.

38 For similar depictions, see Kaul and Kaul (1993:177).

39 Elders could not recollect all the words of the song, but I recorded the following fragments: "They [the raiders] were being chased away from the *phu* (high pasture) of Taru and then they sought shelter in Phyang *phu* and they littered the *phu* with shits all around. The Baltis were being driven away by Hindustani army from Taru *phu* and they were caught in Phyang *phu*. Hail to Ladakh! Hail to Colonel Hari Chand, hail to Colonel Prithi Chand!" In the song, Major Hari Chand is later referred to as Colonel, a title he earned during the war. The song's reference to the soiling of the *phu* alludes to the raiders' excessive consumption of meat. This song was also diffused on the local radio.

Epigraphs: The first is from Verma (1998:136–38). The term Nunnu as employed by many members of the Indian army is the colloquial and affectionate term for Ladakhis. Most commonly encountered as *nono*, it means "younger brother" in Ladakhi. Its use by an Indian officer could be seen as an expression of the paternalistic attitude Indian officers had toward their local tribal troops, which furthered colonial attitudes toward Ladakhis. However, as discussed in this chapter, it also highlights a qualitative shift in the conception of the Ladakhis' attitude from pre- to post-Independence. The second is from Reach Ladakh (2014).

1   For a detailed analysis of the incident, see Saint-Mézard (2013).

2   During the 1950s, India and China conducted discussions over the control of Aksai Chin. For China, the territory's strategic importance was that it provided a link between Xinjiang and Tibet, circumventing the impenetrable Kunlun massif. The Chinese claim refuted the McMahon line, which had established the frontier between India and Tibet at the beginning of the twentieth century. Despite the unsettled issue of sovereignty over Tibet, the Chinese built a segment of the Tibet-Xinjiang highway in Aksai-Chin in 1956–57. As this part of Ladakh was poorly monitored by the Indian authorities, the construction of the road came to be known only months later (Fisher et al. 1963:86).

3   For instance, the suggestion that the heavy military presence in Ladakh produces "an uncomfortable comparison with Chinese-occupied Lhasa" (Margolis 2001:69) is a proposition I have never encountered among Ladakhis.

4   If the events of the Partition in Ladakh left open wounds among a generation of Ladakhis, the violence that took place in the region was much smaller in scale than in other parts of India. Ladakh's Buddhist and Muslim communities notably escaped the atrocities and communal violence that occurred in mass proportions elsewhere in North India, particularly in the Punjab and the western parts of Jammu and Kashmir. In parts of Jammu and Kashmir, a number of Muslim and Hindu communities were exiled and, in some cases, decimated during the Partition (Bose 2003). Today, in parts of Jammu and Kashmir state, the presence of the Indian state is seen by many as an occupying force, and civilian-military relationships differ markedly from those in Ladakh. On this, see Anjum and Varma (2010), Bhan and Trisal (2017), S. Mathur (2016), Robinson (2013), and Varma (2016).

5   By "military agents" I am referring to the fact that members of the Indian army were, given the context of state formation, acting as representatives of the Indian state in their discourse and in their attitude toward the local population.

6   In fact, the political discontent that prevails in the region tends to be focused on the administration of Jammu and Kashmir state (see van Beek 1996) and not on the regional geopolitical context.

7   The Brogpa are an ethnic group who live in a number of villages along the Line of Control (see note 6 in chapter 2 for more details).

8   Although geographically, in Ladakh, the Karakoram constitutes the north-ernmost border, the belief that the Himalayas, which stretch for 2,400 kilometers across the north of the subcontinent, constitute a natural border for India has long prevailed (see, e.g., Census of India 1931, cited in van Beek 1996:111–12).

9   Francis Younghusband ([1904] 2009:111), for instance, commenting on his journey across the Zojila pass, the western gateway to Ladakh, reports: "As regards its natural scenery, it would be difficult to find any more dreary-looking country than Ladakh. Its mountains, though lofty, are not grand or rugged, but resemble a monotonous succession of gigantic cinder-heaps."

10   A consideration for the role of sympathy in state building in the Indian Himalayas also raises questions about what has been called the "bipolar tendencies of sovereign power" (Singh 2012). If vicarious feelings "can spawn hostility as easily as love" (Rutherford 2009:4), this love may not be unconditional. Since their establishment, the Ladakh Scouts (the group of soldiers discussed in this chapter) have contributed to the protection of the Indian state's territorial sovereignty, but only after the Kargil war were the Scouts converted from a battalion to a full-fledged regiment. This raises questions regarding recognition at a higher hierarchical level. More-over, it is only recently that the Indian army has started to recognize the courage of Ladakhi civilians; however, Ladakhis working for the army as porters put their life in danger, and Ladakhi herders living near the borders are often caught in border skirmishes, and also become infor-mants for the army. As has been pointed at, this is something that civil-ian authorities and the political class have largely ignored (Stobdan 2014).

11   These depictions infuse colonial accounts: for instance, in his foreword to Rassul Galwan's autobiography ([1923] 2005), Francis Younghusband compares Ladakhi men to "young boys." In 1860, British Captain William Henry Knight went on to write that the mere sight of strangers scared Ladakhis, a behavior he attributed to the mistreatment by the Dogra army (Sheikh 1999:342); and in 1928, Major Martin Louis Gompertz ([1928] 2000:66) reflected on the Ladakhis' "cheerfulness"—which he notes with empathy—despite what he felt was their poverty-stricken state.

12   As demonstrated by scholars, the politics of imagining Ladakh and its population has taken many forms, often shifting according to geopoliti-cal circumstances; see Aggarwal (2004), Bray (1998), R. Gupta (2013), and van Beek (1998a, 1998b).

13   Similar language can be found in Prithi Chand's boastful description of recruitment of Ladakhis while touring villages of Sham during the first Indo-Pakistani war: "I injected in them the idea of 'Do or Die' for defending Ladakh," he explains, concluding that "the population of Ladakh responded magnificently" (quoted in Chibber 1998:155).

14   After the Sino-Indian war, the Ladakh Scouts were formally raised by amalgamating the 7th and 14th J&K Militia Battalions, which included many Ladakhis who had joined the army during the first war with Pakistan.

15   These officers come from various regiments and are attached to the Scouts for a period of two years.

16   Noteworthy also is the fact that the Pakistani intrusion that triggered the war was discovered by a local Ladakhi while he was grazing his sheep on pastureland that overlooks the India-Pakistan border. His intervention also received extensive media coverage (see Bahn 2014, chapter 3).

17   The Census of India does not publish data on employment in the army, so figures on the number of people thus employed in Ladakh remain elusive. According to *Ladags Melong* (2005:28–29), a local Ladakhi magazine, in 2005 there was about six thousand Ladakhis enrolled in the Ladakh Scouts regiment. But these numbers do not account for employment in paramilitary groups and work carried out for the army (for instance porterage) without enrollment.

18   Employment for women in the army remains very limited.

19   Kashmir, which has been divided between India and Pakistan since the partition of India, remains a disputed territory, claimed as a whole by each. The 1980s saw violent confrontations in Kashmir with the rise of separatist movements (the demands being either autonomy from India or from Pakistan, or for India to cease its claim to Kashmir). A high degree of suspicion by the Indian state for the local population still prevails.

20   Aggarwal (2004:218–19), however, notes this this long-prevailing situation started to change after the Kargil war (1999), when the Indian army realized it needed more personnel who could cope with the high mountains.

21   Issues of representation about Ladakhi identity have been carefully examined by Martijn van Beek, who notes that being Ladakhi does not rest on a stable and homogenous "set of characteristics, forms, idioms, or practice" (van Beek 2003:286). On the various publics and forms of national consciousness in Ladakh, see also Aggarwal (2004), Bertelsen (1996), R. Gupta (2013), and Srinivas (1998).

22   As Samuel (1993:182) notes, the invocation would sometimes be accompanied by the act of placing a stone onto a *lha tho*, one of the stone cairns that frequently mark the highest points of passes and that serve as the abodes of mountain deities.

23   In its casualness, this encounter with the military epitomizes the nature of my interactions with authorities throughout my fieldwork. As I explained

in the introduction, my fieldwork was conducted under a research visa, which set the territorial limits of my research. During my fieldwork, on two occasions I was asked to report to the Jammu and Kashmir Police, in one case because a foreigner seen within the Line of Control had been mistaken for me, the second time because a foreigner made use of the Indian army's satellite network with his own device. Being nearly the only foreigner in the region during the winter, I was the first person to be considered as a potential suspect. My host family in Leh was also subjected to a series of visits when I had to renew my research visa, and I again had to report to the police in order to explain the nature of my research project. In each of these cases, the agents I dealt with were very cordial, and the authorities never interfered with my research in the field. This situation contrasts with what researchers and journalists working in other parts of Jammu and Kashmir state have experienced, in Kashmir in particular (see S. Mathur 2016:157–58 for a review). This illustrates not only the high variability in liberty and access to information within the state of Jammu and Kashmir, but also the fact that interactions with the authorities around Leh are more relaxed than they are in zones that are directly contiguous with the lines of control.

24 The ibex (Siberian ibex, *Capra sibirica*) is described as a "holy animal" (Vohra 1982:91). According to Kaplanian (1983), the ibex is a symbol of the prehistoric and idyllic age of nomadism during which a blurred boundary distinguished the gods from humans. Ibexes are featured in petroglyphs throughout Ladakh. The animal is also prominent in many religious ceremonies: ibex figurines are fashioned from dough and ibex horns are placed on the *lha to*.

25 See Beer (2003) for an analysis of the key symbols of Tibetan Buddhism.

26 This sense of duty takes various guises and materializes in a range of proactive actions that are not limited to enrollment in the army. For instance, mothers' associations voluntarily send food for journeys in the mountains to soldiers at the front, as was the case during the Kargil war of 1999.

27 On this, see the preface by Kushok Bakula Rinpoche in the memoirs of Colonel Chewang Rinchen, *A Legend in His Own Time* (Verma 1998). The lama is careful to highlight the paradox of his intervention: "I wondered at the apparent irony of the choice: requesting a Buddhist monk for writing the Foreword to a Colonel's biography." But he dismisses the irony by referring to "Ahimsa," one of the tenets of Buddhism, citing the following verses, which display a strong notion of sacrifice:

> Victory goes to him
> who fights with complete fearlessness
> and is utterly devoid of hatred

1   The perception of an excess of rainfall may be influenced by a cloudburst that took place in 2010 and brought massive floods into Ladakh. Cloudbursts are, however, not unknown to Ladakhis; they are part of their oral history and they are also mentioned in colonial accounts (see, e.g., Cunningham [1854] 2005:99–111). But many Ladakhis believe that their frequency has increased in the past decade.

2   The gods called *yul lha* (sometimes referred to as "gods of the mountains" or "territorial gods") serve as regulators and guardians of the life of a given space (Kapstein 2006:208). According to Dollfus (1997) and Riaboff (1996), contrary to what has been noted in other Tibetan areas, the *yul lha* of Ladakh are not necessarily associated with mountains but can simply be the gods of a given territory.

3   The *lu* are spirits of the underworld. They like humid, swampy places, which are rich sources of water holes. The *lu* are said to be quite capricious: if they are kept happy, water and wealth will be plenty, if they are polluted or injured, they will respond fiercely by withholding water or sending disease, in particular skin diseases (Samuel 1993:162). For discussions of the *lu* in Ladakh, see Butcher (2013a), Dollfus (2003), Mills (2003), and Sonam Wangchok (2009).

4   Beliefs and interpretations about local gods vary in Tibetan Buddhist areas. Nebesky-Wojkowitz ([1956] 1993:475, 228) defines the *gzhi bdag* (literally, "foundation lord") as "country gods," who are supposed to rule particular places or provinces. He records a list of *gzhi bdag*, a number of which he associates with rivers, lakes, and ridges, but none with glaciers; he nevertheless states that "most seem to be personifications of mountains" (226). Working in Ladakh, Pascale Dollfus (1996:6) translates *gzhi bdag* in French as "maître du lieu," or "master of the place"; when writing in English, she renders the term as "owner of the land," a formulation I often heard myself in Ladakh.

5   Tibetan religious life, though dominated by the traditions of Buddhism and Bon religion, embraces a broad range of beliefs and practices, including those performed by specialist practitioners that appear to be orthodox Buddhism. These elements of religious life—which Stein (1972:191) terms "nameless religion"—involve the participation of religious elites and the general population alike. The practices of the said nameless religion are centered on the "cults of local divinities and spirits, the harmony or conflict between human and the invisible forces with which they must interact" (Kapstein 2006:205–6).

6   This interrelatedness is rooted in the Tibetan Buddhist cosmology, according to which the world is divided between sentient beings, namely all

living beings having the ability to experience and to suffer, and the non-sentient environment, or the "receptacle-world" (Kapstein 2013:7).

7   These details include the place in the field where the plow should first open the ground, the direction in which the first furrow should be traced, and who in the village should be the main actors in the ritual, from the plowman, the man guiding the *dzo*, and the sower to the *dzo* themselves.

8   In 1971, 65 percent of the Ladakhi workforce were listed as cultivators. By 2011, that proportion had decreased dramatically to 28 percent (Government of Jammu and Kashmir 2012). This translates into a critical lack of workforce in the villages. See also Dawa (1999).

9   Labbal (2001) reports that in the past, traveling musicians would go from one plow team to another to sing traditional songs in honor of the *dzo*, songs celebrating the beauty of the high pastures that await the beasts upon completion of their work.

10   The *khangpa* (*khangchen*), or "big house," is the ancestral house; it generally goes to the first son. The *khangu* (*khang chung*), or "small house," is a subsidiary house, where offspring live after the partition of the family land. Generally, the *khangpa* and *khangu* are referred to by the same name. A house is intricately linked to the genealogy of an individual family in Ladakh and has a function akin to a surname.

11   See Ripley (1995) on the social and ritual significance of food in Ladakh, including details on the consumption of *chang*.

12   In Leh District, more than 98 percent of households are holders of ration cards from the Public Distribution System (PDS), India's largest poverty-reduction scheme. The cards enable people to buy wheat and rice at subsidized prices, limited to a quota (Dame and Nüsser 2011:188–90).

13   Yamaguchi et al. (2016) study the social drivers of land abandonment in Ladakh.

14   The regular seasonal melt of a glacier should not be confused with glacier retreat. Every year, a glacier will melt in spring and summer, but the mass a glacier loses during these seasons is normally regenerated in the colder months. Glacier retreat implies the insufficient regeneration of melted glacier ice.

15   In Tingmosgang and elsewhere in Sham, both lay people and experts described the *gzhi bdag* as the protector of the glaciers and the mountains. I have not encountered in other villages of Sham the belief that there is a *gzhi bdag* that dwells in a specific village glacier, nor did I encounter any other instances of the performance of *skyin jug* (a ritual examined in this chapter) for the same deity. It should be noted that although there is a common core to beliefs in local deities, related practices and rituals are known to vary widely in Tibetan Buddhist territories.

16   The link between the behavior of glaciers and the conduct of the laity, who are warned, in order to prevent water stress, not to vex the protector of the glaciers, can also be found in a song from Himachal Pradesh state, on the India-Tibet border, as reported by Giuseppe Tucci (1966:96).

17   Based on astrological calculations, the *lotho* is greatly valued and regularly consulted by farmers. In their study, Angchok and Dubey (2006) found a strong correlation between rainfall predictions from the *lotho* and data collected by the Indian Meteorological Department.

18   Here the expression used to designate the site is *Balti romkhang* (*ro khang*). *Romkhang* is the name used by Shammas to designate a funeral hut. Muslims have different funerary practices, and bodies are buried rather than cremated. Here, *romkhang* is used to designate the site where the bodies of Muslim who died during the war are buried. See Brauen (1982) on funerary customs in Ladakh.

19   In Ladakh, large monasteries appoint individual monks to reside in the village temple and monasteries in order to become *komnyer* (*dgon gnyer*), or caretakers, and village priests. This practice has not been reported elsewhere in the Tibetan region (Samuel 1993:113).

20   Butcher 2013a:111–12; Lopez 1998; Ortner 1998; Pirie 2007:106.

21   Ortner (1989:7–8) reports a similar situation during the course of her research on the establishment of the first celibate Buddhist monasteries among the Sherpa in Nepal. Ortner's informants often directed her to the lama instead of "small" people (lay people), who were not comfortable speaking about religious matters. My lay interlocutors often had the same reaction.

22   *Geshe* is the highest degree in the Gelug school of Buddhism.

23   The professional musicians of Ladakh who traditionally play during rituals belong to the Mons community.

24   Kaplanian (2002) identifies two types of mountain worship in Tibetan areas. The first, more in line with Buddhist orthodoxy, involves pilgrimages to mountains and their circumambulation; the second type, like the *skyin jug* ritual, is associated with territorial deities.

25   *Ju hey* is a salutation, akin to the more common *julley*, or "hello."

26   Ravina Aggarwal, whose work is based in the village of Achinathang in Sham, explains that *skyin jug* comes from the words *skyin*, which means "exchange," and *jug pa*, "to enter" (2004:133). As has been pointed out to me, *skyin jug* may rather derive from *skyid sdug*, which according to Jäschke (1881:26) means "happiness and misery" or "good and ill luck." In Tibetan contexts, *skyid sdug* connotes mutual aid groups (Brox 2016; Gerke 2012; Miller 1956); in a monastic context, *skyid sdug* is defined by Waddell (1895:192) as a form of punishment; as a concept, it is used in religion instructions, such as *skyid sdug lam khyer*, or "turning suffering

and happiness into enlightenment." This latter concept can equally have resonance for the context of a marriage, which is a time of happiness but also, in a patrilocal context, a difficult moment. In Ladakh, I have not encountered *skyid sdug* in any form related to the institution (as is the case for the Tibetan context) and my interlocutors do not link the term *skyin jug* to anything other than a ritual performed by a bride, or the ritual for the glacier of Tingmosgang.

27 *Skyin jug* is referred to as *bag ngus* (tears of the bride) in the Leh region (Aggarwal 2004:137–38). See the work of Kirin Narayan (1986, 1997) in Kangra in the foothills of the Indian Himalayas for *suhag* songs, which have a similarity to *skyin jug*.

28 For a study of wedding ceremonies in Zanskar, see Gutschow (2004:148–56); for Amdo in Tibet, see Skal Bzang Nor Bu and Stuart (1996); for Nepal, see Childs (2004:100–104). All these authors report a phase of overt weeping during the leave-taking ceremony. See also Aggarwal (2004:133–47) for an analysis of *skyin jug* in Ladakh.

29 Written sources on folklore songs performed in the Indian Himalayan belt are limited, but a number of those available use glacier imagery; see Dinnerstein (2013a, 2013b), Tashi Rabgias (1970–2003), Ribbach (1986:65–66), and Tucci (1966:61). Stein (1972:165, 191–229) suggests that folklore of this kind, in being strongly imprinted with morality, is a form of secular religion that may have played a function akin to religion in the past. The oral recitation of lore, according to Stein, reinforces the importance of the wisdom of elders and "lends its sanction to the social and world order, the structure of the environment and that of the group inhabiting it"; this had a protective function, akin to religion, as it was "necessary for upholding the order of the world and society" (198).

30 *Meme* Nyima obviously felt that invoking the nun of Ruth house added credibility to his words. Yet it was not clear to me who she was, for I never managed to meet her. I was told she lived in Leh, but despite many attempts to track her down, I could not find her.

31 As a weather-making ritual, *skyin jug* may be specific to Ladakh, but it echoes diverse Buddhist weather-making ceremonies. See Klein and Sangpo (2007), Nebesky-Wojkowitz ([1956] 1993: chapter 24), and Mills, Huber, and Pedersen (1998).

CHAPTER 5: SEARCHING FOR *ABA STANZIN*

First epigraph: From Lhamo Pemba (1996).

1 The word *kishang* is a composite of dogs (*khi*) and wolves (*shang ku*).

2 I use the expression "livestock herder" as a collective term for people who keep both large livestock (yaks, *dzo*, cows) and small livestock (sheep, goats).

3   Bender 2002; Hirsch and O'Hanlon 1995; Ingold 2000:189–218; Schama 1995; Tilley 1994.

4   It is reported that at some point, the stretch through Baltistan and up to the Mustagh Pass was part of the trading route from Kashmir to Kashgar. When English topographer Henry Haversham Godwin-Austen conducted his survey in 1861, the route was reportedly used only occasionally by a few Baltis who had settled in Yarkand, in order to visit their natal villages (Rizvi 1999:26). An account by French physician François Bernier of his journey in Kashmir in 1664–65 suggests that by the seventeenth century this passage had become extremely challenging due to glacier movement (Bernier 1916:427).

5   *Jhonsa* designates the place where dairy activities take place; it is also sometimes simply referred to as *doksa* (*'brog sa*), or "the high pastures." *Phu* generally designates the upper part of a valley, but people sometimes use it interchangeably with *doksa*, depending on the geography of a village.

6   The situation remains critical. Between 2014 and 2017 two women and a young girl met violent ends through attacks by dogs in different parts of Ladakh.

7   Dollfus (2012 :257–58) describes the different techniques used to track predators in the Changthang area of Ladakh, which are still practiced despite the awareness that they go against the law. In contrast, in Sham, these practices have by and large been abandoned.

8   See Fox et al. (1994) regarding wildlife conservation in Ladakh.

9   Another aspect of this disconnection is the difficulties youths sometimes have in expressing themselves in the colloquial language of their native village as a consequence of prolonged stays outside. In conversations, Ladakhi youths sometimes expressed discomfort at becoming the subject of attention for their lack of familiarity with local idioms.

10 Although never as prevalent as polyandry, polygyny was in the past a widely accepted form of union in Ladakh.

11 See Dollfus and Labbal (2009b) for a detailed study of the units of the land and the mountains in Ladakh.

12 See Osmaston (1994) on crop plants and their uses in Ladakh; and Pordié (2007) on the medical practice of the *amchi* in Ladakh.

13 Classified as vulnerable by the International Union for the Conservation of Nature (IUCN), the total population of urial (*Ovis orientalis vignei*) in India was estimated at 1,000 and animals in a study published close to thirty years ago (Fox et al. 1991:167; see also Fox et al. 1994 and Mallon 1983). According to the same study, Ladakh urial populations showed a dramatic decline in previous decades; the animal was particularly negatively affected by military conflicts during the years 1947–62. No recent data on urial populations in Ladakh are available, but the

Jammu and Kashmir Wildlife Protection Act has most likely helped to stem the animal's decimation.

14 For other cases of how animals are channeling human knowledge about the environment, see Habeck (2006) for the Komi reindeer herders of North Russia; Feit (2004) and C. Scott (1996) for the Cree hunters of Canada; and Lauer and Aswani (2009) for the fishermen of the Solomon Islands.

15 See Dollfus (2012) for an account of the gradual abandonment of pastoralist activities in the Changthang area of Ladakh.

16 Ethnographic studies of hunting practices have long pointed to variability in people's ideas about animals, illustrating that hunting is not determined by economic factors alone. See, for example Brightman (1993), C. Scott (1996), and Tanner (1979).

CHAPTER 6: INTIMATE GLACIERS AND AN ETHICS OF CARE

Second epigraph: From Lhamo Pemba (1996).

1 Several studies have been conducted by anthropologists in Hemis Shukpachan, providing rich details about village life, ritual practices, and household organization; see Day (1989), Dollfus (1989), and Phylactou (1989). However, none of these studies discuss the glacier of the village or engagement with its high pasture.

2 Addressing the question from the perspective of climate, Huber and Pedersen maintain that traditionally in Tibet, the state of the natural environment, as an outcome of weather conditions, is seen as a reflection of the social world and its code of proper conduct; "nature and society were conceived to interact, thereby creating a 'moral climate'" (1997:587–88).

3 Farmers of Sham generally link the source of water in their village to one main glacier and not multiple ones.

4 These names are, however, increasingly forgotten today, as many elders lament.

5 The connection Thubten Jigme Norbu evokes is not unique to those who grew up among glaciers. Amazement at the magnificence of the high mountain peaks is a recurring impression in the accounts of foreign explorers of the Himalayas as well. In his epic travelogue *The Heart of a Continent*, British colonel Francis Younghusband was particularly expressive about the transformative effect of the mountains ([1904] 2009:111). Reflecting on the only time he saw K2, he wrote: "It appeared to rise in an almost perfect cone, but to an inconceivable height. We were quite close under it—perhaps not a dozen miles from the summit—and here on the northern side, where it is literally clothed in glacier, there must have been from fourteen to sixteen thousand feet [4,875 m] of solid ice. It was one of those sights which impress a man forever and produces a permanent

effect upon the mind—a lasting sense of the greatness and grandeur of Nature's works—which he can never lose or forget" (94).

6   For an extensive study of deities in the Buddhist Himalayas, see Nebesky-Wojkowitz ([1956] 1993).

7   Many Ladakhis believe that meritorious people can gaze into these lakes and see the Potala palace and other images of Tibet. On sacred lakes in Ladakh, see also Sonam Wangchok (2009:277–28); for Tibet, Buffetrille (1998).

8   As modes of dwelling, these beliefs and narratives reveal how glacial areas and the high pastures complicate dichotomous conceptions of places, which in Tibetan Buddhist scholarship often oppose sacred and profane, uninhabited and inhabited.

9   Carey 2010; Cruikshank 2005; O'Reilly 2016; Orlove 2005.

10   On evil spirits in Ladakh, see Dollfus and Labbal (2009b:91). Although no herders reported the presence of evil beings other than the *smanmo* in the mountains during the course of our conversations, the presence of these beings has been mapped in different parts of the Himalayas and can be found in various places other than mountains. See Samuel (1993:166–67, 265–67).

11   On ritual practices oriented toward landscape deities in Ladakh, see Mills (2003).

12   The expression is sometimes encountered as *laduk poktches* (*la sdug pog ches*; literally, "to be hurtfully hit by the pass").

13   There are doubts about the historicity of Guru Padmasambhava's travel to Tibet and Ladakh, which is dated to the seventh or eighth century. In Ladakh, the traces of Guru Padmasambhava are to be found in many places, whether in stone imprints or in meditation retreats, normally far from the villages, toward the high pastures.

14   See Butcher (2013a) for an account of contemporary flood prevention through the construction of chorten.

15   In her account of Sherpas working on Mount Everest, Sherry Ortner (1999:127) explains how from the perspective of their religious beliefs, mountaineering has always been questionable, the mountains being the abode of the gods. In order to mediate the risk factor associated with work in this treacherous environment, Sherpas have sought to gain religious protection, for displeasing a god could have fatal consequences. This implies making the proper offerings, but also behaving appropriately by avoiding profane acts and pollution in the mountains, such as killing animals or burning garbage. Villagers in Hemis Shukpachan insist that no specific ritual is required in order to trek to the high pasture and that no deities need to be propitiated to reach close to the glacier, whether one is a local or a foreigner. However, although the risks involved in a trek to Shali Kangri are by no means comparable to those entailed in

climbing Everest (in part because the journey does not entail climbing the glacier itself), there is clearly a sense that wisdom and a careful attitude are requirements. This points to the variability of beliefs relating to glaciers throughout the Himalayas and even Ladakh. As we saw in chapter 4, for example, the glaciers of Tingmosgang are said to be protected by a *zhidak*, which is not the case for Hemis Shukpachan. And in Zanskar, for instance, herders at Pensi-la believe that foreigners conducting scientific studies of the Drang Drung glacier irritate the deities, thus generating bad weather (see my 2016 article, "At the Foot of the Zanskari Glacier," http://glacierhub.org/2016/09/27/at-the-foot-of-a-zanskari-glacier).

CONCLUSION

Epigraph: From Lhamo Pemba (1996).

REFERENCES

Abram, Simone. 2014. "The Time It Takes: Temporalities of Planning." *Journal of the Royal Anthropological Institute* 20 (S1): 129–47.

Aggarwal, Ravina. 2004. *Beyond Lines of Control: Performance and Politics on the Disputed Borders of Ladakh, India.* Durham, NC: Duke University Press.

Aggarwal, Ravina, and Mona Bhan. 2009. "'Disarming Violence': Development, Democracy, and Security on the Borders of India." *Journal of Asian Studies* 68 (2): 519–42.

Akbar, M. K. 1999. *Kargil: Cross Border Terrorism.* Delhi: Mittal Publications.

Angchok, D., and V. K. Dubey. 2006. "Traditional Method of Rainfall Prediction through Almanacs in Ladakh." *Indian Journal of Traditional Knowledge* 5 (1): 145–50.

Angeles, Leonara C., and Rebecca Tarbotton. 2001. "Local Transformation through Global Connection: Women's Assets and Environmental Activism for Sustainable Agriculture In Ladakh, India." *Women's Studies Quarterly* 29:99–115.

Anjum, Aaliya, and Saiba Varma. 2010. "Curfewed in Kashmir: Voices from the Valley." *Economic and Political Weekly* 45 (35): 10–14.

Antze, Paul, and Michael Lambek, eds. 1996. *Tense Past: Cultural Essays in Trauma and Memory.* New York: Routledge.

Appadurai, Arjun. 1996. *Modernity at Large: Cultural Dimensions of Globalization.* Minneapolis: University of Minnesota Press.

Asad, Talal. 1993. *Genealogies of Religion: Discipline and Reasons of Power in Christianity and Islam.* Baltimore: Johns Hopkins University Press.

Asboe, Walter. 1947. "Farmers and Farming In Ladakh (Tibetan Kashmir)." *Journal of the Royal Central Asian Society* 34 (2): 186–92.

Barnes, Jessica, et al. 2013. "Contribution of Anthropology to the Study of Climate Change." *Nature Climate Change* 3:541–44.

Basso, Keith H. 1996. *Wisdom Sits in Places: Landscape and Language among the Western Apache.* Albuquerque: University of New Mexico Press.

Bastien, Edward W. 1983. *Tibetan Buddhism: Cycles of Interdependence*. Madison: South Asia Area Center, University of Wisconsin. DVD.

Bauer, Kenneth Michael. 2004. *High Frontiers: Dolpo and the Changing World of Himalayan Pastoralists*. New York: Columbia University Press.

Bear, Laura. 2014. "For Labour: Ajeet's Accident and the Ethics of Technological Fixes in Time." *Journal of the Royal Anthropological Institute* 20 (S1): 71–88.

———. 2015. *Navigating Austerity: Currents of Debt along a South Asian River*. Stanford, CA: Stanford University Press

Beek, Martjin van. 1996. "Identity Fetishism and the Art of Representation: The Long Struggle for Regional Autonomy in Ladakh." PhD diss., Cornell University.

———. 1998a. "True Patriots: Justifying Autonomy for Ladakh." *Himalaya: The Journal of the Association for Nepal and Himalayan Studies* 18 (1): 35–46.

———. 1998b. "Worlds Apart: Autobiographies of Two Ladakhi Caravaneers Compared." *Focaal* 32:55–69.

———. 2001. "Public Secrets, Conscious Amnesia, the Celebration of Autonomy for Ladakh." In *States of Imagination: Ethnographic Explorations of the Postcolonial State*, edited by T. B. Hansen and F. Stepputat, 365–90. Durham, NC: Duke University Press.

———. 2003. "The Art of Representation." In *Ethnic Revival and Religious Turmoil: Identities and Representations*, edited By Marie Lecomte-Tilouine and Pascale Dollfus, 283–301. New Delhi: Oxford University Press.

———. 2004. "Dangerous Liaisons: Hindu Nationalism and Buddhist Radicalism in Ladakh." In *Religious Radicalism and Security In South Asia*, edited by S. P. Limaye, R. G. Wirsing, and M. Malik, 193–218. Honolulu: Asia-Pacific Center for Security Studies.

Beer, Robert. 2003. *The Handbook of Tibetan Buddhist Symbols* Boston: Shambhala.

Bender, Barbara. 2002. "Time and Landscape." *Current Anthropology* 43 (S4): S103–S112.

Bennett, Jane. 2010. *Vibrant Matter: A Political Ecology of Things*. Durham, NC: Duke University Press.

Berlant, Laurent. 2004. "Introduction: Compassion (and Withholding)." In *Compassion: The Culture and Politics of an Emotion*, edited by Laurent Berlant, 1–14. New York: Routledge.

Bernier, François. 1916. *Travels in the Mogul Empire, A.D. 1656–1668*. Translated by Archibald Constable. London: Oxford University Press.

Berreman, Gerald D. 1964. "*A Study of Polyandry*, by H.R.H. Prince Peter." *Journal of Asian Studies* 23 (2): 321–22.

———. 1980. "Polyandry: Exotic Custom vs. Analytical Concept." *Journal of Comparative Family Studies* 11 (3): 377–84.

Bertelsen, Kristoffer. 1996. *Our Communalized Future: Sustainable Development, Social Identification, and Politics of Representation in Ladakh*. PhD diss., Aarhus University.

———. 1997. "Protestant Buddhism and Social Identification in Ladakh." *Religion, politique et identités en Himalaya,* no. 99: 129–51.

Bhan, Mona. 2008. "Border Practices: Labour and Nationalism among Brogpas of Ladakh." *Contemporary South Asia* 16 (2): 139–57.

———. 2009. "Military Masculinities." In *Recent Research on Ladakh 2009: Papers from the 12th Colloquium of the International Association for Ladakh Studies, Kargil,* edited by Monisha Ahmed and John Bray, 139–44. Leh: International Association for Ladakh Studies.

———. 2014. *Counterinsurgency, Democracy, and the Politics of Identity in India: From Warfare to Welfare?* New York: Routledge.

Bhan, Mona, and Nishita Trisal. 2017. "Fluid Landscapes, Sovereign Nature: Conservation and Counterinsurgency in Indian-Controlled Kashmir." *Critique of Anthropology* 37 (1): 67–92.

Bishop, Peter. 1989. *The Sacred Myth of Shangri-La*. Berkeley: University of California Press.

Bonneuil, Christophe, and Jean-Baptiste Fressoz. 2016. *The Shock of the Anthropocene: The Earth, History and Us*. Translated by David Fernbach. London: Verso.

Borneman, John. 1992. *Belonging in the Two Berlins: Kin, State, Nation*. Cambridge: Cambridge University Press.

Bose, Sumantra. 2003. *Kashmir: Roots of Conflict, Paths to Peace*. Cambridge, MA: Harvard University Press.

Brauen, Martin. 1982. "Death Customs in Ladakh." *Kailash* 9 (4): 319–32.

Bray, John. 1998. "Recent Research on Ladakh: An Introductory Survey." *Himalaya: The Journal of the Association for Nepal and Himalayan Studies* 18 (1): 47–52.

———. 2008. "Corvée Transport Labour in 19th and Early 20th Century Ladakh: A Study in Continuity and Change." In *Modern Ladakh: Anthropological Perspectives on Continuity and Change,* edited by Fernanda Pirie and Martijn van Beek, 43–66. Leiden: Brill Academic Publishers.

Brightman, Robert. 1993. *Grateful Prey: Rock Cree Human-Animal Relationships*. Berkeley: University of California Press.

Brox, Trine. 2016. *Tibetan Democracy: Governance, Leadership, and Conflict In Exile*. New York: I. B. Tauris.

Bubandt, Nils. 2009. "From The Enemy's Point of View: Violence, Empathy, and the Ethnography of Fakes." *Cultural Anthropology* 24 (3): 553–88.

Buch, Elana D. 2013. "Senses of Care: Embodying Inequality and Sustaining Personhood in the Home Care of Older Adults In Chicago." *American Ethnologist* 40 (4): 637–50.

Buddruss, Georg. 1993. "On Artificial Glaciers in the Gilgit Karakorum." *Studien zur Indologie und Iranistik* 18:77–90.

Buffetrille, Katia. 1998. "Reflections on Pilgrimages to Sacred Mountains, Lake, and Caves." In *Pilgrimage in Tibet*, edited by Alex Mckay, 18–34. Richmond, Surrey: Curzon Press.

Butalia, Urvashi. 2000. *The Other Side of Silence: Voices from the Partition of India*. Durham, NC: Duke University Press.

Butcher, Andrea. 2013a. "Keeping the Faith: Divine Protection and Flood Prevention in Modern Buddhist Ladakh." *Worldviews* 17:103–14.

———. 2013b. "Keeping the Faith: An Investigation into the Ways That Tibetan Buddhist Ethics and Practice Inform and Direct Development Activity in Ladakh, North-West India." Phd diss., University of Aberdeen.

Cadena, Marisol de la. 2015. *Earth Beings: Ecologies of Practice across Andean Worlds*. Durham, NC: Duke University Press.

Campbell, Ben. 2013. *Living between Juniper and Palm: Nature, Culture, and Power in the Himalayas*. Oxford: Oxford University Press.

Caplan, Lionel. 1991. "'Bravest of the Brave': Representations of 'The Gurkha' in British Military Writings." *Modern Asian Studies* 25 (3): 571–97.

Carey, Mark. 2010. *In the Shadow of Melting Glaciers: Climate Change and Andean Society*. Oxford: Oxford University Press.

Castree, Noel, et al. 2014. "Changing the Intellectual Climate." *Nature Climate Change* 4:763–68.

Cederlöf, Gunnel, and K. Sivaramakrishnan, eds. 2006. *Ecological Nationalisms: Nature, Livelihoods, and Identities in South Asia*. Seattle: University of Washington Press.

Chakrabarty, Dipesh. 2009. "The Climate of History: Four Theses." *Critical Inquiry* 35 (2): 197–222.

Chibber, Lt. General Manohar Lal. 1998. *Pakistan's Criminal Folly in Kashmir: The Drama of Accession and Rescue of Ladakh*. New Dehli: Manas Publications.

Childs, Geoff. 2004. *Tibetan Diary: From Birth to Death and Beyond in a Himalayan Valley of Nepal*. Berkeley: University of California Press.

Childs, Geoff, Sienna Craig, Cynthia M. Beall, and Buddha Basnyat. 2014. "Depopulating the Himalayan Highlands: Education and Outmigration from Ethnically Tibetan Communities of Nepal." *Mountain Research and Development* 34 (2): 85–94.

Clough, Patricia Ticineto. 2008. "The Affective Turn: Political Economy, Biomedia, and Bodies." *Theory, Culture, and Society* 25:1–22.

Comaroff, Jean, and John L. Comaroff, eds. 2001. *Millennial Capitalism and the Culture of Neoliberalism*. Durham, NC: Duke University Press.

Constable, Nicole. 2009. "The Commodification of Intimacy: Marriage, Sex, and Reproductive Labor." *Annual Review of Anthropology* 38:49–64.

Crate, Susan A., and Mark Nuttall, eds. 2009. *Anthropology and Climate Change: From Encounters to Actions*. Walnut Creek, CA: Left Coast Press.

Crook, John, and Stamati Crook. 1994. "Explaining Tibetan Polyandry: Socio-Cultural, Demographic, and Biological Perspectives." In *Himalayan Buddhist Villages: Environment, Resources, Society, and Religious Life in Zangskar, Ladakh*, edited by J. H. Crook and H. Osmaston, 735–85. Bristol: Bristol Classical Press.

Crook, John, and Tsering Shakya. 1983. "Six Families of Leh." In *Recent Research on Ladakh*, Schriftenreihe Internationales Asienforum, edited by D. Kantowsky and R. Sander. Munich: Weltforum Verlag.

Cruikshank, Julie. 2005. *Do Glaciers Listen? Local Knowledge, Colonial Encounter, and Social Imagination*. Vancouver: UBC Press.

Cunningham, Alexander. [1854] 2005. *Ladakh: Physical, Statistical, and Historical*. Varanasi: Pilgrims Publishing.

Dame, Juliane, and Marcus Nüsser. 2011. "Food Security in High Mountain Regions: Agricultural Production and the Impact of Food Subsidies in Ladakh, Northern India." *Food Security* 3 (2): 179–194.

Das, Veena. 1995. *Critical Events: An Anthropological Perspective on Contemporary India*. Oxford: Oxford University Press.

———. 2007. *Life and Words: Violence and the Descent into the Ordinary*. Berkeley: University of California Press.

Daswani, Girish. 2013. "On Christianity and Ethics: Rupture as Ethical Practice in Ghanaian Pentecostalism." *American Ethnologist* 40 (3): 467–79.

Dave, Naisargi N. 2012. *Queer Activism in India: A Story in the Anthropology of Ethics*. Durham, NC: Duke University Press.

Dawa, Sonam. 1999. "Economic Development of Ladakh: Need for a New Strategy." In *Ladakh: Culture, History, and Development between Himalaya and Karakoram*, edited by Martijn van Beek, Kristoffer Brix Bertelsen, and Poul Pedersen, 369–78. Aarhus: Aarhus University Press.

Day, Sophie. 1989. "Embodying Spirits: Village Oracles and Possession Ritual in Ladakh, North India." PhD diss., London School of Economics.

Deboos, Salomé. 2013. "Religious Fundamentalism in Zanskar, Indian Himalaya." *Himalaya: The Journal of the Association for Nepal and Himalayan Studies* 32 (1): 35–42.

Deleuze, Gilles, and Félix Guattari. 1991. *A Thousand Plateaus: Capitalism and Schizophrenia*. Translated by Brian Massumi. Minneapolis: University of Minnesota Press.

Demenge, Jonathan, Radhika Gupta, and Salomé Deboos. 2013. "Contemporary Publics and Politics in Ladakh." *Himalaya: The Journal of the Association for Nepal and Himalayan Studies* 32 (1): 7–11.

Descola, Philippe. 2013. *Beyond Nature and Culture*. Chicago: University of Chicago Press.

Desjarlais, Robert. 2003. *Sensory Biographies: Lives and Deaths among Nepal's Yolmo Buddhists*. Berkeley: University of California Press.

Diemberger, Hildegard. 2013. "Deciding the Future in the Land of Snow: Tibet as an Arena for Conflicting Forms of Knowledge and Policy." In *The Social Life of Climate Change Models*, edited by Kirsten Hastrup and Martin Skrydstrup, 100–127. New York: Routledge.

Dinnerstein, Noé. 2013a. *Ladakhi Traditional Songs: A Cultural, Musical, and Literary Study*. Diss., City University of New York.

———. 2013b. "Songs, Cultural Representation, and Hybridity in Ladakh." *Himalaya: The Journal of the Association for Nepal and Himalayan Studies* 32 (1): 73–84.

Dollfus, Pascale. 1989. *Lieux de neige et de genévriers*. Paris: Centre National de la Recherche Scientifique.

———. 1996. "Maîtres du sol et dieux du territoire au Ladakh." *Études rurales*, no. 143/144:27–44.

———. 1997. "Mountain Deities among the Nomadic Community of Kharnak (Eastern Ladakh)." In *Ladakh: Culture, History, and Development between Himalaya and Karakoram*, edited by M. van Beek, K. B. Bertelsen, and P. Pedersen, 92–118. Recent Research on Ladakh 8, 8th Colloquium of the International Association for Ladakh Studies, Moesgaard. Aarhus: Aarhus University Press.

———. 2003. "De quelques histoires de klu et de btsan." *Revue d'études tibétaines*, no. 2: 5–39.

———. 2008. "Calcul pour l'ouverture de la bouche de la terre." In *Modern Ladakh: Anthropological Perspectives on Continuity and Change*, edited by Fernanda Pirie and Martijn van Beek, 119–37. Leiden: Brill Academic Publishers.

———. 2012. *Les bergers du Fort Noir: Nomades du Ladakh (Himalaya Occidental)*. Nanterre: Société d'Ethnologie.

Dollfus, Pascale, and Valérie Labbal. 2009a. "A Foray into Ladakhi Place Names." In *Reading Himalayan Landscapes over Time: Environmental Perception, Knowledge, and Practice in Nepal and Ladakh*, edited by Joëlle Smadja, 239–60. Pondicherry: French Institute of Pondicherry

———. 2009b. "Ladakhi Landscape Units." In *Reading Himalayan Landscapes over Time: Environmental Perception, Knowledge, and Practice in Nepal and Ladakh*, edited by Joëlle Smadja, 85–106. Pondicherry: French Institute of Pondicherry

Dyson, Jane. 2015. "Life on the Hoof: Gender, Youth, and the Environment in the Indian Himalayas." *Journal of the Royal Anthropological Institute* 21:49–65.

Faier, Lieba. 2009. *Intimate Encounters: Filipina Women and the Remaking of Rural Japan*. Berkeley: University of California Press.

REFERENCES

Fairhead, James, and Melissa Leach. 1996. *Misreading the African Landscape: Society and Ecology in a Forest-Savanna Mosaic*. Cambridge: Cambridge University Press.

Faizi, Inayatullah. 2007. "Artificial Glacier Grafting: Indigenous Knowledge of the Mountain People of Chitral." *Asia Pacific Mountain Network Bulletin* 8 (1): 6–7.

Feit, Harvey. 2004. "James Bay Crees' Life Projects and Politics: Histories of Place, Animal Partners, and Enduring Relationships." In *Indigenous Peoples and Autonomy*, edited by Mario Blaser, Ravi De Costa, Deborah Mcgregor, and William D. Coleman, 92–110. Vancouver: University of British Columbia Press.

Feuchtwang, Stephan. 2013. "What Is Kinship?" *HAU: Journal of Ethnographic Theory* 3 (2): 281–84.

Fewkes, Jacqueline H. 2009. *Trade and Contemporary Society along the Silk Road: An Ethno-History of Ladakh*. New York: Routledge.

Fisher, Margaret W., Leo E. Rose, and Robert A Huttenback. 1963. *Himalayan Battleground: Sino-Indian Rivalry in Ladakh*. New York: Praeger.

Foucault, Michel. 1976. *Histoire de la sexualité*. Vol. 1: *La volonté de savoir*. Paris: Gallimard.

———. 1994. *Ethics: Subjectivity and Truth*. New York: New Press.

Fox, Joseph L. 1991. "The Mountain Ungulates of Ladakh, India." *Biological Conservation* 58 (2): 167–90.

Fox, Joseph L., Chering Nurbu, Seema Bhatt, and Alok Chandola. 1994. "Wildlife Conservation and Land-Use Changes in the Transhimalayan Region of Ladakh, India." *Mountain Research and Development* 14 (1): 39–60.

Fürer-Haimendorf, Christoph von. 1988. *Himalayan Traders: Life in Highland Nepal*. London: John Murray.

Gagné, Karine. 2016. "Cultivating Ice over Time: On the Idea of Timeless Knowledge and Places in the Himalayas." *Anthropologica* 58 (2): 193–210.

Gagné, Karine, Mattias Borg Rasmussen, and Ben Orlove. 2014. "Glaciers and Society: Attributions, Perceptions, and Valuations." *Wiley Interdisciplinary Reviews: Climate Change* 5 (6): 793–808.

Galwan, Ghulam Rassul. [1923] 2005. *Servant of Sahibs*. New Delhi: Asian Educational Services

Gell, Alfred. 1992. *The Anthropology of Time: Cultural Constructions of Temporal Maps and Images*. Oxford: Berg.

Gellner, David N., ed. 2013. *Borderland Lives in Northern South Asia*. Durham, NC: Duke University Press.

Gerke, Barbara. 2012. *Long Lives and Untimely Deaths*. Leiden: Brill.

Ghosh, Sahana. 2011. "Cross-Border Activities in Everyday Life: The Bengal Borderland." *Contemporary South Asia* 19 (1): 49–60.

Gold, Ann Grodzins. 2001. "Story, Ritual, and Environment in Rajasthan." In *Sacred Landscapes and Cultural Politics: Planting a Tree*, edited by

Philip P. Arnold and Ann Grodzins Gold, 115–137. Aldershot: Ashgate Publishing.

Gold, Ann Grodzins, and Bhoju Ram Gujar. 2002. *In the Time of Trees and Sorrows: Nature, Power, and Memory in Rajasthan*. Durham, NC: Duke University Press.

Goldstein, Melvin. 1978. "Pahari and Tibetan Polyandry Revisited." *Ethnology* 17 (3): 325–37.

Goldstein, Melvin, and Cynthia M. Beall. 1982. "Indirect Modernization and the Status of the Elderly in a Rural Third World Setting." *Journal of Gerontology* 37 (6): 743–48.

Gompertz, Major Martin Louis. 1928. *Magic Ladakh: An Intimate Picture of a Land of Topsy-Turvy Customs and Great Natural Beauty*. London: Seely, Service.

Government of Jammu and Kashmir. 2012. *Statistical Handbook for the Year 2011–2012*. Leh: Ladakh Autonomous Hill Development Council.

Govindrajan, Radhika. 2015a. "'The Goat That Died for Family': Animal Sacrifice and Interspecies Kinship in India's Central Himalayas." *American Ethnologist* 42 (3): 504–19.

———. 2015b. "Monkey Business: Macaque Translocation and the Politics of Belonging in India's Central Himalayas." *Comparative Studies of South Asia, Africa, and the Middle East* 35 (2): 246–62.

———. 2018. *Animal Intimacies: Interspecies Relatedness in India's Central Himalayas*. Chicago: University of Chicago Press.

Greenhouse, Carol J. 1996. *A Moment's Notice: Time Politics across Cultures*. Ithaca, NY: Cornell University Press.

Gregg, Melissa, and Gregory J. Seigworth, eds. 2010. *The Affect Theory Reader*. Durham, NC: Duke University Press.

Gupta, Akhil. 2014. "Authorship, Research Assistants, and the Ethnographic Field." *Ethnography* 15 (3): 394–400.

Gupta, Radhika 2013. "The Importance of Being Ladakhi: Affect and Artifice in Kargil." *Himalaya: The Journal of the Association for Nepal and Himalayan Studies* 32 (1): 43–50.

Gutschow, Kim. 2004. *Being a Buddhist Nun: The Struggle for Enlightenment in the Himalayas*. Cambridge, MA: Harvard University Press.

———. 2006. "The Politics of Being Buddhist in Zangskar: Partition and Today." *India Review* 5 (3–4): 470–98.

Gutschow, Kim, and Niels Gutschow. 1998. "A Landscape Dissolved: Households, Fields, and Irrigation in Rinam, Northwest India." In *Sacred Landscape of the Himalaya*, edited by N. Gutschow, A. Michaels, C. Ramble, and E. Steinkellner, 111–36. Vienna: Austrian Academy of Sciences Press.

Gyatso, Janet. 1987. "Down with the Demoness: Reflections on a Feminine Ground in Tibet." In *Feminine Ground*, edited by J. D. Willis, 33–51. Ithaca, NY: Snow Lion.

REFERENCES

Habeck, Joachim Otto. 2006. "Experience, Movements, and Mobility: Komi Reindeer Herders' Perception of the Environment." *Nomadic Peoples* 30 (2): 123–41.

Haberman, David L. 2013. *People Trees: Worship of Trees in Northern India.* Oxford: Oxford University Press.

Halkias, Georgios T. 2013. "Early Kingdoms of the Himalaya: The Political and Cultural History of the Region to 1700." In *Himalaya: The Exploration and Conquest of the Greatest Mountains on Earth,* edited by Philip Parker. London: Conway.

Haraway, Donna J. 1989. *Primate Visions: Gender, Race, and Nature in the World of Modern Science.* New York: Routledge.

———. 1991. *Simians, Cyborgs, and Women: The Reinvention of Nature.* New York: Routledge.

———. 2008. *When Species Meet.* Minneapolis: University of Minnesota Press.

———. 2015. "Anthropocene, Capitalocene, Plantationocene, Chthulucene: Making Kin." *Environmental Humanities* 6:159–65.

Hardt, Michael. 1999. "Affective Labor." *Boundary 2* 26 (2): 89–100.

Harms, Erik, Shafqat Hussain, Sasha Newell, Charles Piot, Louisa Schien, Sara Shneiderman, Terence Turner, and Juan Zhang. 2014. "Remote and Edgy: New Takes on Old Anthropological Themes." *HAU: Journal of Ethnography Theory* 4 (1): 361–81.

Harvey, David. 1989. *The Condition of Postmodernity: An Enquiry into the Origins of Cultural Change.* Cambridge: Blackwell.

Harvey, Peter. 2000. *An Introduction to Buddhist Ethics.* Cambridge: Cambridge University Press.

Hay, Katherine Eve. 1997. "Gender, Modernization, and Change in Ladakh, India." MA thesis, Norman Paterson School of International Affairs, Carleton University.

Herzfeld, Michael. 2005. *Cultural Intimacy: Social Poetics in the Nation-State.* New York: Routledge.

Hirsch, Eric. 1995. "Landscape: Between Place and Space." In *The Anthropology of Landscape: Perspectives on Place and Space,* edited by Eric Hirsch and Michael O'Hanlon, 1–30. Oxford: Oxford University Press.

Hirsch, Eric, and Michael O'Hanlon, eds. 1995. *The Anthropology of Landscape: Perspectives on Place and Space.* Oxford: Oxford University Press.

Hirsch, Eric, and Charles Stewart. 2005. "Introduction: Ethnographies of Historicity." *History and Anthropology* 16 (3): 261–74.

Hodges, Matt. 2010. "The Time of the Interval: Historicity, Modernity, and Epoch in Rural France." *American Ethnologist* 37 (1): 115–31.

Hopkins, Benjamin D. 2015. "The Frontier Crimes Regulation and Frontier Governmentality." *Journal of Asian Studies* 7 (2): 369–89.

Huber, Toni. 1997. "Green Tibetans: A Brief Social History." In *Tibetan Culture in the Diaspora,* edited by F. J. Korom, 103–19. Vienna: Austrian Academy of Sciences Press.

Huber, Toni, and Poul Pedersen. 1997. "Meteorological Knowledge and Environmental Ideas in Traditional and Modern Societies: The Case of Tibet." *Journal of the Royal Anthropological Institute* 3 (3): 577–98.

Hussain, Shafqat. 2015. *Remoteness and Modernity*. New Haven, CT: Yale University Press.

Igoe, Jim, and Beth Croucher. 2007. "Conservation, Commerce, and Communities: The Story of Community-Based Wildlife Management Areas in Tanzania's Northern Tourist Circuit." *Conservation and Society* 5 (4): 534–61.

Ingold, Tim. 2000. *The Perception of the Environment: Essays on Livelihood, Dwelling, and Skill*. London: Routledge.

———. 2007. *Lines: A Brief History*. New York: Routledge.

———. 2011. *Being Alive: Essays on Movement, Knowledge and Description*. London: Routledge.

Israr-ud-Din. 1960. "A Social Geography of Chitral State." MSc thesis, King's College, University of London.

Jacobs, Nancy J. 2001. "The Great Bophuthatswana Donkey Massacre: Discourse on the Ass and the Politics of Class and Grass." *American Historical Review* 106 (2): 485–507.

Jäschke, Heinrich August. 1881. *A Tibetan-English Dictionary with Special References to the Prevailing Dialects*. London: Secretary of State for India in Council

Jina, Prem Singh. 1994. *Tourism in Ladakh Himalaya*. Delhi: Indus.

Jones, Reece. 2009. "Agents of Exception: Border Security and the Marginalization of Muslims in India." *Environment and Planning D: Society and Space* 27:879–97.

Jordan, Tsering, and Sonika Sen. 2014. "Social Issues of Ageing in Ladakh: The Sociological Insight." *IRC's International Journal of Multidisciplinary Research in Social and Management Sciences* 2 (3): 89–91.

Kaplanian, Patrick. 1983. "Quelques aspects du mythe et des structures mentales au Ladakh." In *Recent Research on Ladakh*, edited by Detlef Kantowsky and Reinhard Sander, 96–106. Munich: Weltforum Verlag.

———. 2002. "De la légende au mythe. Parole, langue et pensée—Book Review of Anne-Marie Blondeau and Ernst Steinkellner, eds., *Reflections of the Mountain: Essays on the History and Social Meaning of the Mountain Cult in Tibet and the Himalaya*." *L'homme* 163 (July–September): 280–83.

Kapstein, Matthew T. 2006. *The Tibetans*. Oxford: Blackwell.

———. 2013. *Tibetan Buddhism: A Very Short Introduction*. Oxford: Oxford University Press.

Karmay, Samten G. 1996. "The Tibetan Cult of Mountain Deities and Its Political Significance." In *Reflections of the Mountain: Essays on the History and Social Meaning of the Mountain Cult in Tibet and the Himalaya*, edited by Anne-Marie Blondeau and Ernst Steinkellner, 59–75. Vienna: Austrian Academy of Sciences Press.

Kaul, Shridhar, and H. N. Kaul. 1992. *Ladakh through the Ages: Towards a New Identity.* New Delhi: Indus.

Khan, Kacho Asfandyar. 1998. *Ancient Wisdom: Sayings and Proverbs of Ladakhi.* Kargil: Kachio.

Khoo, Michael. 1997. "Preliminary Remarks Concerning Solar Observation, Solar Calendars, and Festivals in Ladakh and the Western Himalaya." In *Recent Research on Ladakh 7,* edited by Thierry Dodin and Heinz Räther. Proceedings of the 7th Colloquium of the International Association of Ladakh Studies, Bonn/St.Augustin, June 12–15, 1995. Ulmer Kulturanthropologische Schriften Band 9. Ulm: Ulm University.

Kirksey, S. Eben, and Stefan Helmreich. 2010. "The Emergence of Multispecies Ethnography." *Cultural Anthropology* 25 (4): 545–76.

Klein, Anne C., and Khetsun Sangpo. 2007. "Hail Protection." In *Religion of Tibet in Practice,* abridged ed., edited by Donald S. Lopez Jr., 400–409. Princeton, NJ: Princeton University Press.

Knight, Daniel M., and Charles Stewart. 2016. "Ethnographies of Austerity: Temporality, Crisis, and Affect in Southern Europe." *History and Anthropology* 27 (1): 1–18.

Knight, John, ed. 2000. *Natural Enemies: People-Wildlife Conflicts in Anthropological Perspective.* New York: Routledge.

Kohn, Eduardo. 2013. *How Forests Think: Towards and Anthropology beyond the Human.* Berkeley: University of California Press.

———. 2014. "Toward an Ethical Practice in the Anthropocene." *HAU: Journal of Ethnographic Theory* 4 (1): 459–64.

Labbal, Valérie. 2001. "Un araire du toit du monde." *Techniques et culture* 37:109–24.

Ladags Melong. 2005. "Soldiers of the Soil: The Ladakh Scouts Regiment." *Ladags Melong,* Summer, 28–29.

Laidlaw, James. 2002. "For an Anthropology of Ethics and Freedom." *Journal of the Royal Anthropological Institute* 8 (2): 311–32.

———. 2014. *The Subject of Virtue: An Anthropology of Ethics and Freedom.* Cambridge: Cambridge University Press.

Lambek, Michael. 2008. "Value and Virtue." *Anthropological Theory* 8 (2): 133–57.

———, ed. 2010. *Ordinary Ethics: Anthropology, Language, and Action.* New York: Fordham University Press.

———. 2015. "Chapter 1: Living As If It Mattered." In *Four Lectures on Ethics: Anthropological Perspectives,* by Michael Lambek, Veena Das, Didier Fassin, and Webb Keane. Chicago: HAU Books/University of Chicago Press.

Laszczkowski, Mateusz. 2016. "'Demo Version of a City': Buildings, Affects, and the State in Astana." *Journal of the Royal Anthropological Institute* 22 (1): 148–65.

Latour, Bruno. 1993. *We Have Never Been Modern*. Translated by Catherine Porter. Cambridge, MA: Harvard University Press.

———. 2004. *Politics of Nature: How to Bring the Sciences into Democracy*. Translated by Catherine Porter. Cambridge, MA: Harvard University Press.

———. 2013. "The Anthropocene and the Destruction of the Image of the Globe." Gifford Lectures on Natural Theology, University of Edinburgh, February 25.

———. 2014. "Agency at the Time of the Anthropocene." *New Literary History* 45:1–18.

Lauer, Matthew, and Shankar Aswani. 2009. "Indigenous Ecological Knowledge as Situated Practices: Understanding Fishers' Knowledge in the Western Solomon Islands." *American Anthropologist* 111 (3): 317–29.

Levine, Nancy E. 1988. *The Dynamics of Polyandry: Kinship, Domesticity, and Population on the Tibetan Border*. Chicago: University of Chicago Press.

Lhamo Pemba. 1996. *Tibetan Proverbs*. Dharamsala: Library of Tibetan Works and Archives.

Lopez, Donald S. 1998. *Prisoners of Shangri-La: Tibetan Buddhism and the West*. Chicago: University of Chicago Press.

Lyall, J. B. 1874. *Report of the Land Revenue Settlement of the Kangra District, Panjab, 1865–72*. Lahore: Central Jail Press.

MacDonald, Alexander, ed. 1997. *Mandala and Landscape*. New Delhi: D.K. Printworld (P) Ltd.

Mahmood, Saba. 2005. *The Politics of Piety: The Islamic Revival and the Feminist Subject*. Princeton, NJ: Princeton University Press.

Malhotra, Ashok. 2003. *Trishul: Ladakh and Kargil 1947–1993*. Delhi: Lancer.

Mallon, David. 1983. "The Status of Ladakh Urial *Ovis orientalis vignei* in Ladakh, India." *Biological Conservation* 27 (4): 373–81.

Marchand, Trevor H. J. 2010. "Making Knowledge: Explorations of the Indissoluble Relation between Minds, Bodies, and Environment." *Journal of the Royal Anthropological Institute* 16 (S1): S1–S21.

Marcus, George. 2009. "Introduction." In *Fieldwork Is Not What It Used to Be*, edited by James Faubion and George Marcus, 1–31. Ithaca, NY: Cornell University Press.

Margolis, Eric S. 2001. *War at the Top of the World*. New York: Routledge.

Marin, Andrei, and Fikret Berkes. 2013. "Local People's Accounts of Climate Change: To What Extent Are They Influenced by the Media?" *Wires Climate Change* 4 (1): 1–8.

Massumi, Brian. 2002. *Parables for the Virtual: Movement, Affect, Sensation*. Durham, NC: Duke University Press.

Mathur, Nayanika. 2013. "Naturalizing the Himalaya-as-Border in Uttarakhand." In *Borderland Lives in Northern South Asia*, edited by David N. Gellner, 72–93. Durham, NC: Duke University Press.

———. 2014. "The Reign of Terror of the Big Cat: Bureaucracy and the Mediation of Social Times in the Indian Himalaya." *Journal of the Royal Anthropological Institute* 20 (S1): 148–65.

———. 2015. "A 'Remote' Town in the Indian Himalaya." *Modern Asian Studies* 49 (2): 365–92.

Mathur, Shubh. 2016. *The Human Toll of the Kashmir Conflict: Grief and Courage in a South Asian Borderland.* New York: Palgrave Macmillan.

Mazzarella, William. 2009. "Affect: What Is It Good For?" In *Enchantments of Modernity: Empire, Nation, Globalization,* edited by Saurabh Dube, 291–309. New Delhi: Routledge.

McGranahan, Carole. 2010. *Arrested Histories: Tibet, the CIA, and Memories of a Forgotten War.* Durham, NC: Duke University Press.

Michaels, Alex. 1998. "The Sacredness of (Himalayan) Landscapes." In *Sacred Landscape of the Himalaya,* edited by A. Michaels, N. Gutschow, C. Ramble, and E. Steinkellner, 13–18. Vienna: Austrian Academy of Sciences Press.

Middleton, Townsend, and Jason Cons. 2014. "Coming to Terms: Reinserting Research Assistants into Ethnography's Past and Present." *Ethnography* 15 (3): 279–90.

Miller, Beatrice D. 1956. "Ganye and Kidu: Two Formalized Systems of Mutual Aid among the Tibetans." *Southwestern Journal of Anthropology* 12 (2): 157–70.

Mills, Martin. 2003. *Identity, Ritual, and State in Tibetan Buddhism: The Foundations of Authority in Gelukpa Monasticism.* Abingdon: Routledge Curzon.

———. 2006. "The Silence in Between: Governmentality and the Academic Voice in Tibetan Diaspora Studies." In *Critical Journeys: The Making of Anthropologists,* edited by Geert De Neve and Maya Unnithan-Kumar, 192–205. London: Ashgate.

———. 2007. "Re-Assessing the Supine Demonness: Royal Buddhist Geomancy in the Srong Btsan Sgam Po Mythology." *Journal of the International Association of Tibetan Studies* 3:1–47.

Mills, Martin, Toni Huber, and Poul Pedersen. 1998. "Ecological Knowledge in Tibet." *Journal of the Royal Anthropological Institute* 4 (4): 783–86.

Mitchell, W. J. T. 2002. *Landscape and Power.* Chicago: University of Chicago Press.

Moore, Donald S. 2005. *Suffering for Territory: Race, Place, and Power in Zimbabwe.* Durham, NC: Duke University Press.

Muehlebach, Andrea. 2011. "On Affective Labor in Post-Fordist Italy." *Cultural Anthropology* 26 (1): 59–82.

———. 2012. *The Moral Neoliberal: Welfare and Citizenship In Italy.* Chicago: University of Chicago Press.

Munn, Nancy D. 1992. "The Cultural Anthropology of Time: A Critical Essay." *Annual Review of Anthropology* 21:93–123.

Nadasdy, Paul. 2003. *Hunters and Bureaucrats: Power, Knowledge, and Aboriginal-State Relations in the Southwest Yukon*. Vancouver: UBC Press.

———. 2007. "The Gift in the Animal: The Ontology of Hunting and Human-Animal Sociality." *American Ethnologist* 34 (1): 25–43.

Narayan, Kirin. 1986. "Birds on a Branch: Girlfriends and Wedding Songs in Kangra." *Ethos* 14 (1): 47–75.

———. 1997. "Singing from Separation: Women's Voices in and about Kangra Folk Songs." *Oral Tradition* 12 (1): 25–53.

Navaro-Yashin, Yael. 2012. *The Make-Believe Space: Affective Geography in a Postwar Polity*. Durham, NC: Duke University Press.

Nazarea, Virginia D. 2006. "Local Knowledge and Memory in Biodiversity Conservation." *Annual Review of Anthropology* 35:317–35.

Nebesky-Wojkowitz, René de. [1956] 1993. *Oracles and Demons of Tibet: The Cult and Iconography of the Tibetan Protective Deities*. Delhi: Book Faith India.

Nora, Pierre. 1989. "Between Memory and History: *Les Lieux de Mémoire*." *Representations* 26 (special issue: Memory and Counter-Memory): 7–24.

Norbu, Thubten Jigme, and Heinrich Harrer. 1986. *Tibet Is My Country: Autobiography of Thubten Jigme Norbu, Brother of the Dalai Lama, as Told to Heinrich Harrer*. London: Wisdom Publications.

Norbu, Thubten Jigme, and Colin Turnbull. 1972. *Tibet: Its History, Religion, and People*. Harmondsworth: Penguin Books.

Ogden, Laura A. 2011. *Swamplife: People, Gators, and Mangroves Entangled in the Everglades*. Minneapolis: University of Minnesota Press.

Ong, Aihwa. 1999. *Flexible Citizenship: The Cultural Logics of Transnationality*. Durham, NC: Duke University Press.

O'Reilly, Jessica. 2016. "Sensing the Ice: Field Science, Models, and Expert Intimacy with Knowledge." *Journal of the Royal Anthropological Institute* 22 (S1): 27–45.

Orlove, Ben. 2002. *Lines in the Water: Nature and Culture at Lake Titicaca*. Berkeley: University of California Press.

———. 2005. "Human Adaptation to Climate Change: A Review of Three Historical Cases and Some General Perspectives." *Environmental Science and Policy* 8 (6): 589–600.

———. 2009. "The Past, the Present, and Some Possible Futures of Adaptation." In *Adapting to Climate Change: Thresholds, Values, Governance*, edited by W. Neil Adger, Irene Lorenzoni, and Karen L. O'Brien, 131–63. Cambridge: Cambridge University Press.

Orlove, Ben, Ellen Wiegandt, and Brian H. Luckman, eds. 2008. *Darkening Peaks: Glacier Retreat, Science, and Society*. Berkeley: University of California Press.

Ortner, Sherry B. 1989. *High Religion: A Cultural and Political History of Sherpa Buddhism.* Princeton, NJ: Princeton University Press.

———. 1998. "The Case of the Disappearing Shamans, or No Individualism, No Relationalism." In *Selves in Time and Place: Identities, Experience, and History in Nepal,* edited by Debra Skinner, Alfred Pach, and Dorothy C. Holland, 239–67. Lanham: Rowman & Littlefield.

———. 1999. *Life and Death on Mt. Everest: Sherpas and Himalayan Mountaineering.* Princeton, NJ: Princeton University Press.

Osmaston, Henry. 1994. "The Farming System." In *Himalayan Buddhist Villages: Environment, Resources, Society, and Religious Life in Zangskar, Ladakh,* edited by John H. Crook and Henry Osmaston, 139–198. Bristol: Bristol Classical Press.

Pálsson, Gísli. 1994. "Enskilment at Sea." *Man* 29 (4): 901–27.

Pandey, Gyanendra. 2002. *Remembering Partition: Violence, Nationalism, and History in India.* Cambridge: Cambridge University Press.

Pandian, Anand. 2008. "Tradition in Fragments: Inherited Forms and Fractures in the Ethics of South India." *American Ethnologist* 35 (3): 466–80.

———. 2009. *Crooked Stalks: Cultivating Virtue in South India.* Durham, NC: Duke University Press.

Pandian, Anand, and Daud Ali. 2010. "Introduction." In *Ethical Life in South Asia,* edited by Anand Pandian and Daud Ali, 1–20. Bloomington: Indiana University Press.

Parreñas, Rheana "Juno" Salazar. 2012. "Producing Affect: Transnational Volunteerism in a Malaysian Orangutan Rehabilitation Center." *American Ethnologist* 39 (4): 673–87.

Peter, Prince of Greece and Denmark. 1963. *A Study of Polyandry.* The Hague: Mouton.

Phylactou, Maria. 1989. "Household Organisation and Marriage in Ladakh Indian Himalaya." PhD thesis, University of London.

Piketty, Thomas. 2014. *Capital in the Twenty-First Century.* Translated by Arthur Goldhammer. Cambridge, MA: Harvard University Press.

Pirie, Fernanda. 2007. *Peace and Conflict in Ladakh: The Construction of a Fragile Web of Order.* Leiden: Brill.

Pommaret, Françoise. 2004. "Rituels aux divinités locales de Kheng 'Bu Li (Bhoutan Central)." *Revue d'études tibétaines* 6:60–77.

Pordié, Laurent. 2007. "Buddhism in the Everyday Medical Practice of the Ladakhi 'Amchi.'" *Indian Anthropologist* 37 (1): 93–116.

Povinelli, Elizabeth. 2006. *The Empire of Love.* Durham, NC: Duke University Press.

Prasad, S. N., and Dharm Pal. [1978] 2005. *History of Operations in Jammu and Kashmir (1947–48).* Dehradun: Natraj Publishers and the Mininstry of Defence, Government of India.

Rabgias, Tashi, ed. 1970–2003. *Ladvags gyi yul glu* [*Ladakhi Folk Songs*]. Leh: Jammu and Kashmir Academy of Art, Culture, and Languages.

Raffles, Hugh. 2002. "Intimate Knowledge." *International Social Science Journal* 54 (173): 325–35.

Ramble, Charles. 1990. "How Buddhists Are Buddhist Communities: The Construction of Tradition in Two Lamaist Villages." *Journal of the Anthropological Society of Oxford* 21 (2): 85–197.

———. 1997. "The Creation of the Bon Mountain of Kongpo." In *Mandala and Landscape*, edited by Alexander W. Macdonald, 133–232. New Delhi: D.K. Printworld (P) Ltd.

Ramsay, H. L. 1890. *Western Tibet: A Practical Dictionary of the Language and Customs of the Districts Included in the Ladakh Wazarat*. Lahore: N.p..

Rasmussen, Mattias Borg. 2015. *Andean Waterways: Resource Politics in Highland Peru*. Seattle: University of Washington Press.

Reach Ladakh. 2014. "In Conversation with Col. Sonam Wangchuk." www .Reachladakh.Com/In-Conversation-With-Col-Sonam-Wangchuk/2466 .html.

Renu, Dr. 2013. "Battle Tested Ladakh Scouts." *Asia Defence News* 8 (5): 30–31.

Riaboff, Isabelle. 1996. "gZhon nu mdung lag, Mountain God of Zanskar: A Regional-Scale Divinity and Its Cult's Territorial Ordering." In *Reflections of the Mountain: Essays on the History and Social Meaning of the Mountain Cult in Tibet and the Himalaya*, edited by Anne-Marie Blondeau and Ernst Steinkellner, 23–37. Vienna: Austrian Academy of Sciences Press.

Ribbach, Samuel. 1986. *Culture and Society in Ladakh*. Translated by John Bray. Delhi: Ess Ess Publications.

Ripley, Abby. 1995. "Food as Ritual." In *Recent Research on Ladakh 4 & 5: Proceedings of the Fourth and Fifth International Colloquia on Ladakh*, edited by Henry Osmaston and Philip Denwood, 165–77. Delhi: Motilal Banarsidass.

Rizvi, Janet. 1996. *Ladakh: Crossroads of High Asia*. Oxford: Oxford University Press.

———. 1999. *Trans-Himalayan Caravans: Merchant Princes and Peasant Traders in Ladakh*. Oxford: Oxford University Press.

Robinson, Cabeiri Debergh. 2013. *Body of Victim, Body of Warrior: Refugee Families and the Making of Kashmiri Jihadists*. Berkeley: University of California Press.

Rogers, Douglas. 2009. *The Old Faith and the Russian Land: A Historical Ethnography of Ethics in the Urals*. Ithaca, NY: Cornell University Press

Rutherford, Danilyn. 2009. "Sympathy, State Building, and the Experience of Empire." *Cultural Anthropology* 24 (1): 1–32.

———. 2016. "Affect Theory and the Empirical." *Annual Review of Anthropology* 45:285–300.

Sahlins, Marshall. 2013. *What Kinship Is—and Is Not*. Chicago: University of Chicago Press.

Saint-Mézard, Isabelle. 2013. "L'incident frontalier du printemps 2013: Un essai d'interprétation des relations sino-indiennes." *Hérodote* 3 (150): 132–49.

Salick, Jan, Anja Byg, and Kenneth Bauer. 2012. "Contemporary Tibetan Cosmology of Climate Change." *Journal for the Study of Religion, Nature, and Culture* 6 (4): 447–76.

Samuel, Geoffrey. 1993. *Civilized Shamans: Buddhism in Tibetan Societies.* Washington, DC: Smithsonian Institution Press.

Schama, Simon. 1995. *Landscape and Memory.* New York: Vintage Books.

Scott, Colin. 1996. "Science for the West, Myth for the Rest? The Case of James Bay Cree Knowledge Construction." In *Naked Science: Anthropological Inquiry into Boundaries, Power, and Knowledge,* edited by Laura Nader, 69–86. London: Routledge.

Scott, James C. 2009. *The Art of Not Being Governed: An Anarchist History of Upland Southeast Asia.* New Haven, CT: Yale University Press.

Sen, L. P. 1969. *Slender Was the Thread: Kashmir Confrontation 1947–48.* New Delhi: Orient Longmans.

Shaheen, F. A., M. H. Wani, S. A. Wani, and Chewang Norphel. 2013. "Climate Change Impact in Cold Arid Desert of North-Western Himalaya: Community-Based Adaptations and Mitigations." In *Knowledge Systems of Societies for Adaptation and Mitigation of Impacts of Climate Change,* edited by Sunil Nautiyal et al., 239–56. Berlin: Springer.

Sheeraza Ladakh. 1998–99. Special edition, vol. 20, nos. 3–4.

Sheikh, Abdul Ghani. 1998–99. "Spyi-lo 1947 nang la-dwags su dmag 'thab byang-pa'i gnas lugs." *Sheeraza Ladakh* 20 (3–4): 184–210.

———. 1999. "Economic Conditions in Ladakh during the Dogra Period." In *Ladakh: Culture, History, and Development between Himalaya and Karakoram,* edited by Martjin van Beek, Kristoffer Bertelsen, and Poul Pederson, 339–49. Aarhus: Aarhus University Press.

Singh, Bhrigupati. 2012. "The Headless Horseman of Central India: Sovereignty at Varying Thresholds of Life." *Cultural Anthropology* 27 (2): 383–407.

Sitaraman, Srini. 2013. "Deconstructing the PLA's Border Incursion in Eastern Ladakh." *East Asia Forum,* August 9.

Sivaramakrishnan, K. 2015. "Ethics of Nature in Indian Environmental History: A Review Article." *Modern Asian Studies* 49 (4): 1261–1310.

Skal Bzang Nor Bu and Kevin Stuart. 1996. "The Rdo Sbis Tibetan Wedding Ceremonies." *Anthropos* 91 (4–6): 441–55.

Smith, Sara. 2009. "The Domestication of Geopolitics: Buddhist-Muslim Conflict and the Policing of Marriage and the Body in Ladakh, India." *Geopolitics* 14 (2): 197–218.

———. 2011. "'She Says Herself, "I Have No Future"': Love, Fate, and Territory in Leh District, India." *Gender, Place, and Culture: A Journal of Feminist Geography* 18 (4): 455–76.

———. 2012. "Intimate Geopolitics: Religion, Marriage, and Reproductive Bodies In Leh, Ladakh." *Annals of the Association of American Geographers* 102 (6): 1511–28.

———. 2013a. "'In the Past, We Ate from One Plate': Memory and the Border in Leh, Ladakh." *Political Geography* 35:47–59.

———. 2013b. "Intimate Territories and the Experimental Subject in Ladakh, India." *Ethnos*:1-22.

Snellgrove, David L., ed. and trans. 1967. *The Nine Ways of Bon: Excerpts from "gZi-brjid."* London: Oxford University Press.

Sophie, Day. 1989. "Embodying Spirits: Village Oracles and Possession Ritual in Ladakh, North India." PhD thesis, London School of Economics.

Sonam Wangchok. 2009. "Sacred Landscapes in the Nubra Valley." In *Mountains, Monasteries, and Mosques: Recent Research on Ladakh and the Western Himalaya*, edited by John Bray and Elena De Rossi Filibeck, 271–83. Pisa: Fabrizio Serra Editore.

Srinivas, Smriti. 1998. *The Mouths of People, the Voice of God: Buddhists and Muslims in a Frontier Community of Ladakh.* Delhi: Oxford University Press.

Stafford, Charles, ed. 2013. *Ordinary Ethics in China.* London: Bloomsbury.

Steffen, Will, Jacques Grinevald, Paul Crutzen, and John McNeill. 2011. "The Anthropocene: Conceptual and Historical Perspectives." *Philosophical Transactions of the Royal Society A* 369:842–67.

Stein, Rolf Alfred. 1972. *Tibetan Civilization.* Stanford, CA: Stanford University Press.

Stewart, Kathleen. 2007. *Ordinary Affects.* Durham, NC: Duke University Press.

Stobdan, P. 2014. "Army's Ingenious Frontier Diplomacy." *Institute for Defence Studies and Analysis*, February 5.

Stoler, Ann Laura. 2002. *Carnal Knowledge and Imperial Power: Race and the Intimate in Colonial Rule.* Berkeley: University of California Press.

———. 2004. "Affective States." In *A Companion to the Anthropology of Politics*, edited by Daniel Nugent and Joan Vincent, 4–20. Oxford: Blackwell.

Strathern, Marilyn. 1988. *The Gender of the Gift: Problems with Women and Problems with Society in Melanesia.* Berkeley: University of California Press.

Tambiah, Stanley J. 1997. "Friends, Neighbors, Enemies, Strangers: Aggressor and Victim in Civilian Ethnic Riots." *Social Science and Medicine* 45 (8): 1177–88.

Tanner, Adrian. 1979. *Bringing Home Animals: Religious Ideology and Mode of Production of the Mistassini Cree Hunters.* New York: St. Martin's Press.

Thornton, Thomas F. 2008. *Being and Place among The Tlingit.* Seattle: Unversity of Washington Press.

Thrift, Nigel. 2008. *Non-Representational Theory: Space, Politics, Affect.* London: Routledge.

Tilley, Christopher. 1994. *A Phenomenology of Landscape*. Oxford: Berg.

Tsering Samphel. 1997. "Zorawar Singh, Tshultim Nyima, and the End of the Ladakhi Monarchy." In *Recent Research on Ladakh 7*, edited by Thierry Dodin and Heinz Räther, 421–26. Proceedings of the 7th Colloquium of the International Association of Ladakh Studies, Bonn/St.Augustin, June 12–15, 1995. Ulmer Kulturanthropologische Schriften Band 9. Ulm: Ulm University.

Tsering, Tashi. 2008. "The Green Primitives of the Himalayas Revisited." *Diaspora, Indigenous, and Minority Education: Studies of Migration, Integration, Equity, and Cultural Survival* 2 (4): 295–301.

Tsewang Rigzin. 2005. "The Impact of the Army In Ladakh." *Ladags Melong* (Summer): 24–30.

Tsing, Anna. 1993. *In the Realm of the Diamond Queen: Marginality in an Out-of-the-Way Place*. Princeton, NJ: Princeton University Press.

———. 2015. *The Mushroom at the End of the World: On the Possibility of Life in Capitalist Ruins*. Princeton, NJ: Princeton University Press.

Tsing, Anna, Nils Bubandt, Elaine Gan, and Heather Anne Swanson, eds. 2017. *Arts of Living on A Damaged Planet: Ghosts and Monsters of the Anthropocene*. Minneapolis: University of Minnesota Press.

Tucci, Giuseppe. 1966. *Tibetan Folk Songs from Gyantse and Western Tibet*. Ascona, Switz.: Artibus Asiae.

Tveiten, Ingvar Nørstegård. 2007. "Glacier Growing—A Local Response to Water Scarcity in Baltistan and Gilgit, Pakistan." Thesis, Noragric, Department of International Environment and Development Studies, Norwegian University of Life Science (UMB), Ås.

Vanquaille, Amandus, and Hilde Vets. 1998. "Lamayuru: The Symbolic Architecture of Light." In *Sacred Landscape of the Himalaya*, edited by N. Gutschow, A. Michaels, C. Ramble, and E. Steinkellner, 85–94. Vienna: Austrian Academy of Sciences Press.

Varma, Saiba. 2016. "Love in the Time of Occupation: Reveries, Longing, and Intoxication in Kashmir." *American Ethologist* 43 (1): 50–62.

Vasan, Sudha. 2017. "Being Ladakhi, Being Indian: Identity Formation, Culture, and Community." *Economic and Political Weekly* 52 (14): 43–49.

Verma, Virendra. 1998. *A Legend in His Own Time (Chewang Rinchen)*. Dehradun: Young India Publications.

Vohra, Rohit. 1982. "Ethnographic Notes on the Buddhist Dards of Ladakh: The Brog-Pā." *Zeitschrift für Ethnologie* 107 (1): 69–94.

Waddell, Laurence Austine. 1895. *The Buddhism of Tibet*. London: W. H. Allen & Co.

Willow, Anna J. 2011. "Conceiving Kakipitatapitmok: The Political Landscape of Anishinaabe Anticlearcutting Activism." *American Anthropologist* 113 (2): 262–76.

Yamaguchi, Takayoshi, Sonam Ngodup, Mitsuhiro Nose, and Shinya Takeda. 2016. "Community-Scale Analysis of the Farmland Abandonment Occurrence Process in the Mountain Region of Ladakh, India." *Journal of Land Use Science* 11 (4): 401–16.

Younghusband, Francis. [1904] 2009. *The Heart of a Continent: A Narrative of Travels in Manchuria, across the Gobi Desert, through the Himalayas, the Pamirs, and Chitral, 1884–1894*. Memphis, TN: General Books.

Zigon, Jarrett. 2007. "Moral Breakdown and the Ethical Demand: A Theoretical Framework for an Anthropology of Moralities." *Anthropological Theory* 7 (2): 131–50.

INDEX

affect: concept of, 11–13, 144; embodiment, intimacy of the everyday, and, 10–13; glaciers and, 145; intimacy as, 146; knowledge and, 121; mountains and, 144; proximity, affective, 146; state and affective sentiments, relation between, 78–80

affective labor, 75–76, 88–89

Aggarwal, Ravina, 75–76, 186n21, 188n23, 192n20, 196n26

agriculture. *See* farming and agrarian life

Aksai Chin, 19*map*, 73, 190n2

altitude sickness, 77, 79, 149, 157, 160

amchi (*am chi*) (traditional medicine practitioners), 79, 127

Ang, 3, 92. *See also* Tingmosgang

Angchok, D., 196n17

Anglo-Sikh war, first (1845–46), 181n8

animals. *See* herders and human-animal relations

Anthropocene, 17–18

Antze, Paul, 50

Asboe, Walter, 36

astrologers (*onpo*) (*dbon po*), 41, 96, 97, 97*fig.*, 99

bag ngus (tears of the bride), 197n27

Balti, 48, 70, 105, 136, 184n2, 185n18

Balti romkhang (*rok hang*), 196n18

Balti rong (Purig), 55–56, 185n14, 189n31

Baltistan, 19*map*; 1948 war and, 50, 52, 59, 64; about, 184n2; British colonialism and, 80, 181n9; glacier grafting in, 177n3; Muslims from Sham migrating to, 61–62; trading with, 32, 55–56. *See also* Indo-Pakistani war, first (1948)

Basgo, 59–66

Basso, Keith, 139

Bauer, Kenneth Michael, 42

Bear, Laura, 144

Beek, Martijn van, 192n21

begar (forced labor), 34, 51–52

Bender, Barbara, 159

Bernier, François, 198n4

Betso Shi Sa ("place of the dead calf"), 37

Bhan, Mona, 75–76, 88, 184n6, 186n22, 189n35

"bipolar tendencies of sovereign power," 191n10

Bishop, Peter, 77

boarding schools, 126–27, 151

bodily experiences. *See* embodiment

bokhari (wood stove), 22, 23–24, 31

border area, Ladakh as: affect and, 12–13; affective labor, military, 75–76, 88–89; Depsang Incident (2013), 73, 85–86; Dogra rule and,

border area (*continued*)
180n28; Ladakh Scouts and,
82–84, 86–89; militarization and
Ladakh as strategic region, 72–75,
165; pervasiveness of the border,
85–90; production of border subjectivity, 76; reimagining of frontier and people, 80–85; sentinel
citizenship and, 76–77, 84–85, 89;
sympathetic identification, 77–
80, 82–83
British rule: Commercial Treaty
(1870), 181n9; Dogra and, 34, 43–
44; first Anglo-Sikh war (1845–
46), 181n8; Ladakh described as
harsh and dangerous, 77; narratives of mountain peoples, 80, 83;
State Forces and, 184n3
Brogpa community, 75, 88, 184n6,
191n7
Buddhism: "green," 178n5; orthodox
trend in, 107, 125–26. *See also*
cosmology, Tibetan Buddhist;
religion
Buddruss, Georg, 178n3
buses, 116–17
Butcher, Andrea, 179n11, 189n37

Caplan, Lionel, 83
care, notion of, 7–8. *See also* ethics
of care
cattle herding. *See* herders and
human-animal relations
Chand, Hari, 59–61, 70, 132, 189n39
Chand, Khushal, 66, 68–69, 189n36
Chand, Prithi, 53, 57, 70, 79, 189n39,
192n13
Chewang Rinchen, 193n27
Chibber, Manohar Lal, 57
Chiktan, 55–56, 62, 185n14
Chil Chil Kangri, 147
Childs, Geoff, xiv, 111–12, 135
China, 73, 80, 85–86, 89, 165, 190n2

chortens, 66–69, 69*fig.*, 153, 153*fig.*,
189n37
citizenship, sentinel, 76–77, 84–85, 89
climate change, 16–18, 104, 140
Commercial Treaty (1870), 181n9
communalism: military and, 84; 1948
war and, 56, 58, 61–62, 188nn24–25;
Partition violence and, 190n4
community work obligations, moral
obligation toward, 43
companion species, 132
construction as bad omen, 43
cosmology, Tibetan Buddhist: and
ethics of care, 8–10; interrelatedness in, 136, 194n6; life cycle in,
26, 65–66, 135–36, 169; territorial
deities, 95–96; verticality, 147, 149
Cruikshank, Julie, 170–71
Cunningham, Alexander, 77

Daulat Beg Oldi, 85
death, attitude toward, 33
debt, 33–35
deities. *See* divine beings and deities
Delek Namgyal, King, 59
Deleuze, Gilles, 179n12
depot duty under Dogra rule, 36
Depsang Incident (2013), 73, 85–86
Desjarlais, Robert, 14, 171
development, 30, 37
Dharamsala, xiii
divine beings and deities: affect and,
12; agriculture and relationship
with territorial deities, 95–96;
avoiding irritating, 147; moral
obligations and, 8; mountains as
dwelling place of, 144; water supply and, 104. *See also* rituals
Dogra rule: attitude towards Ladakhi,
186n21; border status and, 180n28;
capture of Ladakh, 33–34; community cohesion and, 43–44; hardships and violence during, 34–35;

resistance, 36; Treaty of Leh (1842), 181n7

Dollfus, Pascale, 179n9, 194n2, 194n4, 198n7

domestication, 132. *See also* herders and human-animal relations

Dubey, V. K., 196n17

Durkheim, Émile, 7, 93

duty, sense of, 88, 193n26

dzo/dzomo (cow-yak hybrid): bond with, 134–35; farming work by, 92–93, 100, 100*fig.*; honoring of, 96, 195n9; on roof, 40–41. *See also* herders and human-animal relations; livestock

economy: agro-pastoralist, displacement of, 89–90, 93, 126; Chinese incursions as threat to, 85–86; dairy, 122; military and, 84, 89–90; seasonal outmigration and, 31; at time of Indian independence, 32. *See also* farming and agrarian life; herders and human-animal relations

elders: death, attitude toward, 33; ethical dilemmas of, 169; household reorganization and, 44; old age in Tibetan life cycle, 26, 65–66, 169; winter and, 38–41

elders, opinions of. *See specific topics, such as* glaciers

embodiment: affect, intimacy of the everyday, and, 10–13; intensity, embodied, 147; knowledge of the mountains and, 118–23, 142; military, affect, and, 79

epochs, 16

ethics: animals and, 121; anthropological approach to, 7; dilemmas of elders, 169; embodied, 11; experiential dimension of, 142; Kantian vs. Aristotelian, 178n6; morality vs., 8. *See also* moral order; morality

ethics of care: affect, embodiment, everyday intimacy and, 10–13; as embodied ethics, 11; glaciers and, 140–41, 147–48; herding and, 135–36; land, responsibility to, 102; morality and, 7–8, 166–67; as practice, 6–7; terminology and notion of care, 7–8; Tibetan Buddhist cosmology and, 8–10

family, extended, as cooperative unit, 42–43. *See also* household

farming and agrarian life: abandoned fields, 127; Chonpa family and, 98–103; cycles of interdependence, 94–97; filial link, erosion of, 114–15; landscape kinship and, 115; military economy and, 89–90; *sa kha phye* ritual (first plowing), 96–97; season of, 94–95; time regimes and, 93–94; tractor rental and mechanization of, 92–93, 102–3; uncertainty of the present, 115; water concerns and *skyin jug* ritual, 103–7, 109–15, 195n15, 196n26; workforce, lack of, 93

Fewkes, Jacqueline H., 84, 186n21

fieldwork conditions, 23–26

firewood, 31, 35

floods, 60, 153, 182n21, 188n27

folk stories and folk songs, xiii–xiv

Foucault, Michel, 6

fragments of traditions, 10

Frontier Crimes Regulation, 80

Fürer-Haimendorf, Christoph von, 185n16

Galwan, Rassul, 191n11

Gellner, David N., 180n28

Gilgit, 19*map*, 80

Gilgit Scouts, 13, 47–48

glaciers: artificial, 178n4; care and temporality of, 146–48, 159–60; Chil Chil Kangri, 147; embodiment, sense of self, and enskilment in the mountains, 157–59; filial link, erosion of, 114–15; intimate, relational knowledge of, 142–46, 156, 159–60; Kangri Nyingpa ("old glacier") and Kangri Soma ("new glacier"), 5, 103; Kyeri (Tibet), 142–43; lack of knowledge and concern by youth, 139; as lifeblood of Ladakh, 19; monitoring and nurturing, 104; moral selfhood and, 160–62; nurturing practices, 5–6, 170, 177n3; recession of, 5, 6, 18, 94, 136, 140–41, 147, 170–72, 195n14; seasonal melting of, 103, 195n14; *skyin jug* ritual, 104–7, 109–15, 195n15, 196n26; songs, imagery in, 111; trek to Shali Kangri, 139–40, 148–60; water scarcity and, 63, 103–4

goat herding. *See* herders and human-animal relations

Godwin-Austen, Henry Haversham, 198n4

Golden Jubilee, Ladakh Scouts, 87–89

Gompertz, Martin Louis, 36, 191n11

government schools, 126–27

Govindrajan, Radhika, 11

Guattari, Félix, 179n12

Gurkha, 83

Gutschow, Kim, 189n31

hair as status marker, 181n6

Haraway, Donna, 132, 145

Hardt, Michael, 88

Hemis Labrang, 59–61, 64, 188n30

Hemis Shukpachan, 52, 54, 134, 139–40, 151–60

herders and human-animal relations: dairy economy and, 122; domestication and companion species,

132; embodied knowledge of the mountains and, 118–23; glaciers and, 145–46; the herder's journey, 120; knowledge of glaciers and mountains, 143; landscape and, 117–18; predation and predator control, 3–4, 18, 117, 123–26, 128, 132, 133*fig.*; *ra res* daily grazing system, 122–23; reciprocity, interspecies bonds, and attitudes, 134–36; trek to *phu* (high pasture), 121–22, 126–33, 129*fig.*, 131*fig.*; youth views of, 128

high pasture. See *phu*

history vs. memory, 50

Hodges, Matt, 16

household: as cooperative unit, 42; independence of India, impact of, 43; reorganization of, 44; as security net for elders, 39; as site of integration, 29–30

Huber, Toni, 178n5, 199n2

Hume, David, 78

ibex, 87, 193n24

imagination, 12, 79, 80–85

India, independence of (1947): clear distinction between pre- and postindependence India, 36–37; household organization, effect on, 43; Ladakh economy at time of, 32; Partition and, 13, 190n4

Indo-Pakistani war, first (1948): affect and, 12–13; Basgo village and Hemis Labrang massacre, 59–66; bridge at Khalatse and chorten to Major Chand, 66–69, 69*fig.*; conscription system, 53; impacts on Ladakhis, 13–14; invasion of Jammu and Kashmir, 47–48; Kargil allegiance to India, 186n22; Ladakhi support for India, 13, 48–49, 53–54; Leh, defense of,

landscape (*continued*)
changes to, 6; narratives of, 154–56; scholarship on, 117–18; temporality of, 159

lapotse (*la pog ches*) (mountain sickness), 149, 160. *See also* altitude sickness

Latour, Bruno, 80

Leh: defense of, in 1948 war, 13, 49, 50–51, 53, 64, 66–69; description of, 22–23; development and, 37; Dogra rule and, 35; Ladakh Scouts Golden Jubilee, 87–89; march of raiders toward, 48, 55; military aircraft in, 72–73; paid care of homes in, 43; Prithi Chand speech (1948), 57; relocation to, 30, 31; trade route and, 32; war memorial, Hall of Fame Museum, 76*fig.*; in winter, 29, 37

Leh District, 18, 19*map*, 180n22

lha tho (*lha mtho*) (altars to mountain deities), 95*fig.*, 107, 192n22

life cycle in Tibetan Buddhist cosmology, 26, 65–66, 135–36, 169

Likir, 123–24, 126–33

Likir monastery, 107–9, 109*fig.*, 124

Line of Actual Control (China and India), 19*map*, 85–86

Line of Control (Pakistan and India), 19*map*

livestock: domestication and "companion species," 132; raiders' slaughter of, 54–55, 185n10; selling of, 4, 136. *See also* dzo/dzomo; herders and human-animal relations

loneliness, 30, 37, 168–69

Lorimer, D. L. R., 178n3

lotho (*le tho*) (almanacs), 105, 196n17

lu (*klu*) (underworld spirits), 95–96, 107, 194n3

Lyall, J. B., 169–70

Maitreya statues, 108, 189n37

Malhotra, Ashok, 53

marriage: polyandry, 42–43, 55, 182nn22–23, 183nn26–29; polygyny, 183n26, 198n10

Massumi, Brian, 147

McMahon line, 190n2

memory vs. history, 50

Mi Duk Sa, 150

Mi K'at Sa, 154–56, 155*fig.*, 159

militarization of the border. *See* border area

military: acceptance of, 74–75; affective labor and, 75–76, 88–89; conscription, 53, 150; economic significance of, 84; Ladakh Scouts, 82–84, 86–89, 191n10, 192n17, 192nn14–15; Leh and military aircraft corridor, 72–73; region as whole and, 165; sympathetic identification, 77–80, 82–83. *See also* Indo-Pakistani war, first

monasteries, 106, 107–9, 109*fig.*, 124, 196n19

moral order: climate change, moral nature of, 16–18; defined, 8; glaciers and, 141; landscape changes and, 6; Pirie on, 179n10; social time and, 14; war as marker of social time and, 13–16

morality: community work obligations and, 43; ethics vs., 8; first Indo-Pakistani war as moral dilemma, 14–15, 49–50; glacier recession "because people have become bad," 140–41; military affective labor and, 88–89; nonreligious sources of, xiii–xv, 166; selfhood, moral, 151, 160–62

mountain imagery, 111–12

mountains, relationship with. *See* glaciers; *phu*

Raffles, Hugh, 146
rainfall, 18, 94, 194n1
Ramble, Charles, xiv
rape accounts during 1948 war, 64, 189nn31–32
Rasmussen, Mattias Borg, 16
reciprocity: care and, 8; divine-human relationships and, 9; glaciers and, 140–41; human-animal, 135
religion: ethics of care and, 9–10; folk or "nameless" religion, 179n11, 194n5; generation gap in knowledge of, 107; glaciers and, 147; morality beyond doctrine, xiii–xv, 166; 1948 war and politics of religious identity, 54–59; pragmatic approach to, xv–xvi, 39; pragmatic use of religious experts, xv–xvi; Sherpas and, 200n15. *See also* cosmology, Tibetan Buddhist; rituals
remoteness, 29–30. *See also* isolation, sense of
Riaboff, Isabelle, 194n2
rituals: glaciers and, 147–48; to placate territorial deities, 96; *sa kha phye* (first plowing), 96–97; *sang (bsangs)* (juniper burning), 39; *skyin jug* (appeal to *zhidak*), 104–7, 109–15, 195n15, 196n26; *storma (gtor ma)* (exorcist offerings), 52, 184n8
Rizvi, Janet, 185n16, 185n18
Rogers, Douglas, 8
Rutherford, Danilyn, 75, 78–80, 84

sa kha phye ritual (first plowing), 96–97
sadak (sa bdag) (lord of the soil), 95
Sahlins, Marshall, 115
Samuel, Geoffrey, 192n22
sang (bsangs) ritual, 39
Saspol, 54, 64

Saspotse, 117, 123
schools, 126–27, 151
Scott, James, 165
seasonal cycle, 37–38, 121–22, 181n2. *See also* winter
selfhood and sense of self, 62, 149, 151, 158, 160–62, 169
sentinel citizenship, 76–77, 84–85, 89
Shali Kangri, trek to, 139–40, 148–60
Sham, 20*map*; description of, 28; Muslim-Buddhist relations in, 61–62, 187n23; territory of, 177n1. *See also specific locations*
shang dong (wolf pits), 125
shang shang practice, 125
sheep herding. *See* herders and human-animal relations
Sheikh, Abdul Ghani, 187n23
Sherpas, 200n15
sibling position, 42
Singh, A. P., 72
Singh, Gulab, 181n8
Singh, Ranjit, 181n8
Singh, Zorawar, 34, 186n21
Sino-Indian War (1962), 73
Sivaramakrishnan, K., 7
skyin jug ritual, 104–7, 109–15, 195n15, 196n26
smanmo (sman mo) (female ghosts), 120, 147, 150, 200n10
snow decline, 5
snow leopards, 3–4, 4*fig.*, 18, 123
social time, 15–16, 93
Sonam Wangchuk, 72, 74, 89
songs, 111, 195n9, 197n27, 197n29
spang (damp, grassy area), 127, 152
spies, suspected, 67
Spinoza, Baruch, 179n12
splitness, sense of, 41–42
state, production of, 14, 75–76, 88, 165–66, 167. *See also* border area; military
Stein, Rolf Alfred, 86, 179n11, 194n5

storma (*gtor ma*) rituals, 52, 184n8
summer settlements, 121–22, 132,
 154, 159
sun markers (*nyi tho*), 37
Suru Valley, 32
sympathetic identification, 77–80

Tambiah, Stanley, 186n19
taxation, 34–35
tehsildar (revenue official), 34, 35,
 182n12
temperatures, average, 180n24
temporality. *See* time and
 temporality
Tia, 130, 131
Tibetan Buddhism. *See* cosmology,
 Tibetan Buddhist; religion
time and temporality: agrarian, shift
 in, 103; glaciers and, 146–48, 159–
 60; herding and, 121–22; regimes
 of, 93–94; social, 15–16, 93
Tingmosgang, 3–5, 92–94, 98–115,
 100*fig.*
Tingmosgang monastery, 106
tourism, 85–86, 150, 164, 180n1
tractors, rental of, 93, 102–3
Treaty of Leh (1842), 181n7
tree plantations, 134
tsogspa (*tsogs pa*) (village bodies),
 183n30
Tucci, Giuseppe, 196n16
Turnbull, Colin, 135
Tveiten, Ingvar, 178n3

Ulley, 123
urials (wild mountain sheep), 128,
 198n13

vegetation as resource, 127
vegetation decline, 122
Verma, Virendra, 78, 81
verticality, Buddhist notion of, 147,
 149

virtue, sense of, 166–67
volunteer work, 88

Waddell, Laurence Austine, 196n26
war as marker of social time, 13–16
war memorial, Hall of Fame Museum,
 Leh, 76*fig.*
war with Pakistan. *See* Indo-Pakistani
 war, first
water scarcity: glaciers and, 63, 103–
 4; *skyin jug* ritual, 104–7, 109–15,
 195n15, 196n26; snow decline and,
 5; *zhidak* stubbornness and, 104
weather-making ceremonies, 112–13,
 197n31. See also *skyin jug* ritual
wedding ceremonies, 110, 197n28
Wildlife Protection Act, Jammu and
 Kashmir, 125, 199n13
winter: experience of, 28–31, 37–41;
 farming as preparation for, 92,
 102; fieldwork conditions in, 24–
 25; income, lack of, 180n1; in
 Nyemo, 28; sun markers and, 37
women: economic restructuring,
 impact of, 99; military and, 192n18;
 military husbands and, 99; 1948
 war and, 48, 51–52; *skyin jug*
 poetry and, 110
workforce, lack of: agrarian, 93; pastoral, 126

Yangthang, 123
Younghusband, Francis, 77, 118–19,
 191n9, 191n11, 199n5
yul lha (territorial gods), 95–96,
 194n2

Zangs Lung Nge Go, 150
Zanskar, 122, 132, 201n15
zhidak (*gzhi bdag*) ("foundation
 lords"), 95–96, 104, 107, 110–15,
 194n4, 201n15
Zojila pass, 191n9

CULTURE, PLACE, AND NATURE
Studies in Anthropology and Environment

www.ingramcontent.com/pod-product-compliance
Lightning Source LLC
Chambersburg PA
CBHW031125270326
41929CB00011B/1499